Jefferson Sebastian Andrade Macas
Pedro Javier Zapata Coll

Integrated Management of Whitefly Pest (Paraleyrodes Minei)

Jefferson Sebastian Andrade Macas
Pedro Javier Zapata Coll

Integrated Management of Whitefly Pest (Paraleyrodes Minei)

and Cottony Fly (Aleurothrixus Floccosus) in Navelina Orange

Imprint
Any brand names and product names mentioned in this book are subject to trademark, brand or patent protection and are trademarks or registered trademarks of their respective holders. The use of brand names, product names, common names, trade names, product descriptions etc. even without a particular marking in this work is in no way to be construed to mean that such names may be regarded as unrestricted in respect of trademark and brand protection legislation and could thus be used by anyone.

Cover image: www.ingimage.com

This book is a translation from the original published under ISBN 978-620-3-03099-0.

Publisher:
Sciencia Scripts
is a trademark of
International Book Market Service Ltd., member of OmniScriptum Publishing Group
17 Meldrum Street, Beau Bassin 71504, Mauritius
Printed at: see last page
ISBN: 978-620-3-48498-4

INTEGRATED MANAGEMENT OF WHITE MOSCLE *(Paraleyrodes minei)* AND CLOUDFLAW *(Aleurothrixus floccosus)* PEST IN NAVELINE ORANGE.

AUTHOR: Jefferson Sebastian Andrade Macas

TUTORS: Pedro Javier Zapata Coll

Summary

Citrus production in the Spanish Levante region is an activity of great economic importance and has cultural characteristics related to the area. In recent years, whitefly *(Paraleyrodes minei) and* cotton fly *(Aleurothrixus floccosus)* have caused damage to plant development and fruit destruction. In the absence of environmentally friendly control methods, a trial has been carried out to evaluate the integrated control of whitefly and cotton fly, combining auxiliary fauna with spot treatments.

The critical period for the crop in which the experiment was carried out was from the summer season until harvesting, the study and control of the pests was carried out in July, August and September. Weekly sampling was used to determine the population dynamics of the phytophagous pests and natural enemies.

The auxiliary fauna controlled the whitefly and the cotton fly efficiently. However, with the high temperatures, their population decreased and, in contrast, the pest pressure increased, so a phytosanitary treatment was applied. The systemic action of the pesticide reduced the population to tolerable levels until the populations of auxiliary fauna were replenished to control the pest. This integrated control method for whitefly *(Paraleyrodes minei) and* cotton fly *(Aleurothrixus floccosus)* was found to be effective.

Key words: integrated production, whitefly, cotton fruit fly, fungus.

INTEGRATED MANAGEMENT OF WHITE FLY PEST *(PARALEYRODES MINEI)* AND COTTON FLY *(ALEUROTHRIXUS FLOCCOSUS)* IN NAVELINA ORANGE.

Abstract

Citrus production in the Spanish east is an activity of great economic importance and has cultural characteristics related to the area. In recent years, White Fly *(Paraleyrodes minei)* and Cotton Fly *(Aleurothrixus floccosus)* have caused damage to plant development and fruit destrous. In the absence of environmentally friendly control methods, a trial has been carried out in this work to evaluate the integrated control of White Fly and Cotton Fly, combining auxiliary fauna with timely treatments.

The critical period for the crop in which the experiment was conducted was from the summer season to the collection, study and control of pests was carried out in july, August and September. Weekly sampling determined the dynamics of phytophage populations and natural enemies.

Auxiliary wildlife efficiently controlled the white fly and cotton fly. However, with high temperatures its population declined and in contrast, the pressure of the pest increased so a phytosanitary treatment was applied. The systemic action of phytosanitary plant protection reduced the population to tolerable levels until the uptick in auxiliary fauna populations for control. Thus, this integrated control method for White Fly *(Paraleyrodes minei)* and Cotton Fly *(Aleurothrixus floccosus)* was effective.

Keywords: integrated production, white fly, cotton fly, fungal.

CONTENTS

1. Introduction

The agri-food sector in Spain today is geared towards a social, environmental and food good. The main function is to supply food to meet demand. However, these foods are required to have certain characteristics for the consumer.

Agricultural activity is distributed throughout the territory and has an impact on the environment, playing an important role in the management of natural resources and climate change mitigation. For this reason, integrated management techniques must be developed to obtain better production with the least use of resources.

It is a sector that is resistant to crises, in addition to its strategic importance. The Spanish agri-food industry is the fourth largest agricultural producer in the European Union with 23.2 million cultivated hectares, 12% of the total, behind France (17%), Germany (13%) and Italy (13%).

In 2018, the production of the vegetable branch reached 32 billion euros according to the Ministry of Agriculture, Fisheries and Food, which is 60% of the entire production of the agricultural branch.

1.1. Description of the crop

Origin and distribution

The origin of citrus fruits is located in East Asia, in an area stretching from the southern slopes of the Himalayas to southern China, Indochina, Thailand, Malaysia and Indonesia (Maroto Borrego, 1998).

The oldest known quotation comes from China and belongs to the "Book of History" (5th century BC). It explains how the emperor Ta-Yu (23rd century BC) included among his taxes the delivery of two types of oranges, large and small. This indicates the high value attributed to these species (González-Sicilia, 1968).

Citrus fruits have a long history in Spain. Citrons were introduced by the Romans in the 5th century. Initially citrus fruits were used as ornamental plants and for medicinal purposes, and after the introduction of sweet oranges also for fresh consumption on a very local scale. Commercial plantations and exports of fresh fruit to other European countries started at the end of the 18th century (Abad, 1984).

The cultivation of the mandarin tree *(Citrus reticulata* Blanco) began in 1856 from plant material imported to the Plana de Castellón, probably from Palermo, Genoa and Nice, where its cultivation was already known. The bitter orange and lemon trees were brought to Spain by the Arabs in the 11th century, via Africa and from Arabia (Agustí, 2003).

Finally, the grapefruit *(Citrus paradisi)* was the last citrus fruit to be introduced in Spain. The first plants known to be present were of the "Marsh" variety and were imported by the Estación Naranjera de Levante, in 1910, from the USA (Zaragoza, 1993). (Zaragoza, 1993).

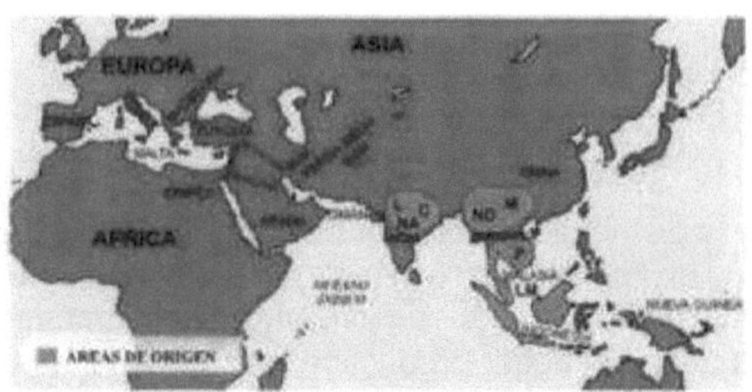

Illustration 1: Map showing the areas where commercial citrus species originated. C: citron, NA: bitter orange, L: lemon, LM: lime, P: grapefruit, ND: sweet orange, M: mandarin (source: Zaragoza et al., 2011).

The orange is a citrus fruit that belongs to the Rutaceae family. The most commercialised sweet oranges in Spain belong mainly to the Navel (Navelina, Washington Navel, Navelate, Lane late) and Blancas (Valencia late) groups (MAGRAMA, 2012).

1. 1. 2. Economic importance of the orange tree

Economic importance in the world

Citrus is an important fruit crop in the world, whose production in 2018 was over 11 million tonnes and 15 million hectares planted according to FAO (Food and Agriculture Organization of the United Nations). It is grown in more than 100 countries, mostly in tropical and subtropical regions in both hemispheres.

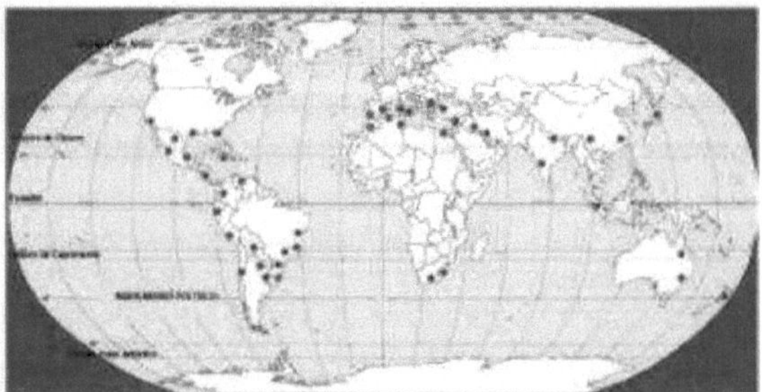

Illustration 2: Distribution of the main citrus-producing areas in the world (Source: Zaragoza et al., 2011).

The main citrus-producing countries worldwide are Brazil and China with around 23 million tonnes, followed by the United States with almost 8 million tonnes, and Mexico with more than 4 million tonnes. Spain is in sixth place.

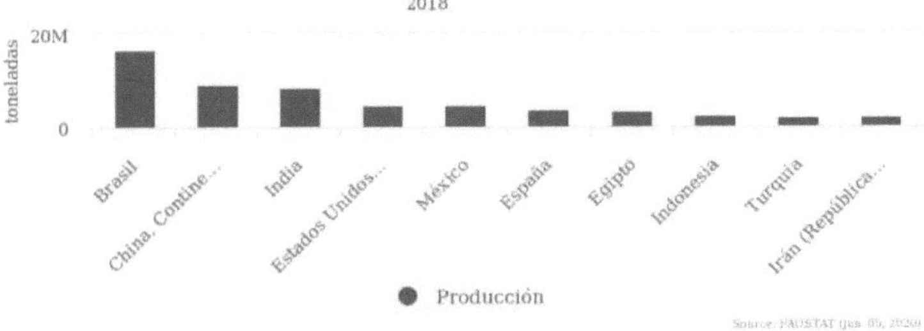

Table 1: Top 10 citrus producers. FAOSTAR 2018.

According to FAO statistics for the year 2018, produced   oranges are the most popular citrus fruit in the world with 75,413,374 tonnes. China is the world's
leading producer with around 17,800,000 tonnes, accounting for 49.3% of global production.

Cultivo	Producción (Toneladas)
Naranjas	75413374
Limones y limas	19368838
Tangerinas, mandarinas, clementinas y satsumas	34393430
Toronja y pomelo	9374739

Table 2: World citrus production in 2019. FAOSTAT 2020.

In terms of international trade, the world production of oranges is traded on international markets as fresh fruit, but most of the production, especially in countries with the largest planted area, such as Brazil or the United States, is destined for the industrial manufacture of juices (Bravo, 2014).

The volume of world imports of fresh oranges reached 7.1 million tonnes in 2017 with a value of USD 5,662 million in 2017 (FAO, 2017). The European Union accounts for almost 50% of the global import volume. The Netherlands has the largest share in the world import volume with 8.2%, followed by France (7.3%), Germany (6.5%), Russian Federation (6.0%) and Saudi Arabia (5.4%) (FAO, 2017).

Imports of the top 5 importers of Oranges The world export volume of fresh oranges is close to 7 million tonnes in 2017 with a value that reached USD 5.1 billion in 2017 (FAO, 2017), experiencing an increase of 14.4% in the period 2007-2017.

Graph 1: Imports of the top 5 importers of oranges. FAOSTAR 2017.

Exportaciones de los 5 exportadores principales de Naranjas

Graph 2: Exports of the top 5 orange exporters. FAOSTAR 2017.

Spain is the world's leading exporter of fresh oranges with a 23.4% share. It is followed by South Africa with a share of 16.9%; Egypt, with (15.8%); the United States (8.5%) and Turkey (5.6%) (FAO, 2017). According to the FAO, Spain exports more than half of its production.

Economic importance in Spain
Spain ranks 6th with 139,626 hectares cultivated and approximately 3.6 tonnes of oranges (FAOSTAT, 2017), which is an important producer of oranges at the international level.

There are citrus plantations along the Mediterranean coast, in the provinces of Tarragona, Castellón, Valencia, Murcia, Almería and Málaga, in the Guadalquivir river valley in the provinces of Córdoba and Seville and on the Atlantic coast in the province of Huelva (Illustration 3).

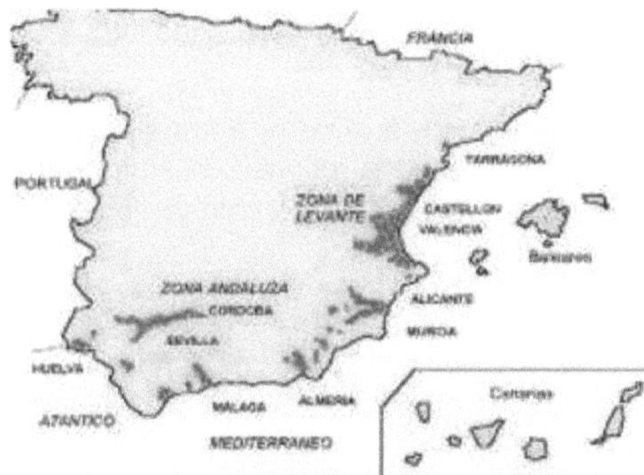

Illustration 3: Main citrus-producing areas in Spain. Zaragoza (2011)

In Spain, 7,528,310 tonnes of citrus fruits were produced in 2018 according to MAPAMA's Statistical Yearbook. Sweet oranges accounted for 51.9% of total production, mandarins and lemons 31.8% and 14.9%, respectively.

The surface area of cultivated orange trees is concentrated in the Valencia Community, with 50.3%, followed by Andalusia (40.7%) and Murcia (5.4%). This represents 96.4% of Spain's cultivated area of oranges (MAPAMA, 2019).

Spanish citriculture, particularly Valencian citriculture, has a strong exporting vocation for a better price. Oranges are mainly exported for fresh consumption with high quality standards as they do not have the presence of pests and high or no phytosanitary residues, which is why production is carried out with integrated pest control.

Morphological characteristics

Botanical classification

The orange tree has several commercial interests of citrus fruits from the agronomic point of view whose species belong to:

- *Rutaceae* family
- Subfamily *Aurantioideas.*
- Division *Embriophyta Siphonogama*
- Subdivision *Angiospermae*
- Class *Dicotyledonae,*
- Subclass *Rosidae,*
- Superorder *Rutanae*
- *Routine* Order
- Orange (*Citrus sinensis*)

There is currently no unanimous agreement on citrus taxonomy, although two classifications are most commonly used: Swingle's, which considers 16 species, and Tanaka's, which considers 162 (Zaragoza, 2011).

Tree

They are characterised by being evergreen trees or shrubs ranging in height from 5 to 16 metres depending on the species, variety, rootstock and soil and climatic conditions, although they are currently cultivated in the form of dwarf varieties which are easier to farm and more productive.

In the adult state, they are normally formed by a trunk that branches out profusely to a height of 60-80 cm, and forms a rounded, bushy crown. The economically useful lifespan is 30-40 years, although there are trees that are more than 100 years old (Zaragoza, 2011).

Illustration 4: Navelina orange tree. Source: IVIA.

Flower
They are usually gathered in inflorescences, generally in the form of corymbs, although more rarely they appear isolated. They have 5 petals and numerous stamens (Agustí, et al., 2010).

Illustration 5: Orange blossom Citrus sinensis.

Fruit
It is a typical berry called hesperidium, where the flavedo, albedo and endocarp, inside which the seeds are found, can be clearly distinguished (IVIA, 2010). The usual size is between 3 and 10 centimetres, although there are species that reach 30 cm (Agustí, et al., 2010).

A characteristic feature of the genus is the presence of an essential oil in all organs of the plant, which gives it its characteristic odour. Citrus fruits are one of the most popular foods worldwide, due to their taste, vitamin C content and high nutrient/calorie ratio (Kimball, 2002).

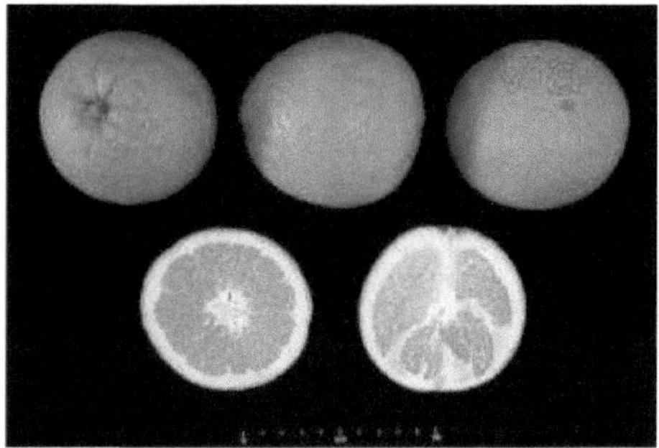

Illustration 6: Orange fruit Navelina variety. Source: IVIA.

The harvesting period for citrus fruits varies according to the variety cultivated, although in general they are harvested from autumn until the end of winter. I would like to highlight the early variety Navelina, the subject of the study, whose fruit is an intense orange colour, with high levels of acidity that it maintains until the end of the harvesting period.

Illustration 7: Orange harvesting periods. Source: IVIA.

1. 2. Crop requirements

1. 2. 1. Climatic requirements

Citrus fruits are temperate climate plants because the most important climatic variable in determining vegetative development, flowering, fruit set and fruit quality is temperature. Temperatures of 25 to 30oC are considered optimal for photosynthetic activity and temperatures of 35oC or higher reduce it (Agustí, 2003).

Most pests occur in spring-summer with optimal temperatures for photosynthetic activity and the development of phytophagous pests. For this reason, integrated production control is necessary to obtain the best yields and the quality demanded by the markets.

In the Spanish Levante region, the main risk for the production is frost when temperatures fall below 2°C for more than two hours. The damage caused by frost may vary from the loss of quality of part of the fruit to the total loss of the crop.

Tolerance to low temperatures varies according to the following factors: species or genus, variety, vegetative state of the tree, age of the plant, health status, presence of micro- and macronutrient deficiencies, rootstock and growing conditions (Soler, 2006).

1. 2. 2. Soil requirements
Citrus can grow under very different soil conditions, from stony, very poor soils to heavy clay soils. However, this is not an indication that the crop is equally adapted to all soil types. While they are able to thrive in soils without conditions, they do so at the cost of their vegetative development and production (Arevalo, 2013).

They thrive best in deep sandy soils and loamy soils, provided that light, temperature, mineral elements and water are not limiting. On the other hand, impermeable and very clayey soils hinder their growth. When the proportion of clay is higher than 50%, root growth is severely restricted (Arévalo, 2013).

The most suitable soils for growing citrus fruits are those with an equal proportion of fine elements (clays and silts) and coarse elements (sands), with which they associate the qualities of well-drained soils, but with adequate retention of the soil's aqueous solution, which guarantees good nutrition for the trees (Agustí, 2003).

In terms of soil permeability, suitable soils with medium permeability, between 10 and 30 cm/h, should be avoided. Soils with a permeability of more than 40 cm/h, which are incapable of retaining water, or less than 5 cm/h, which are easily waterlogged, should be avoided.

As far as the pH reaction of the soil is concerned, it is not an important factor in itself in citrus cultivation. In fact, it is very common to find optimal harvests, in terms of quantity and quality, in soils with pH between 5 (moderately acid) and 8.5 (moderately alkaline).

The active part of the roots is located in the superficial layers of the soil, at a depth of between 0.5 and 0.7 metres, which must allow them to develop, without obstacles, down to the subsoil (Agustí, 2003). Therefore, in addition to considering the characteristics of the soil, it is also necessary to consider those of the subsoil.

The ideal soil for citrus cultivation in general is one that meets these conditions and the limestone content, expressed as calcium carbonate, must be between 10 and 20% (Soler, 2006).

Soil conditions and root development determine the size of the tree stand and is closely related to its production. When soil conditions inhibit root development, production is reduced. Therefore, factors that can damage the root system and reduce its density (waterlogging, excessive fertilisation, closely spaced tillage, etc.) in turn reduce yield (Agustí, 2003).

1. 2. 3. Water requirements

Citrus trees are very demanding both in terms of quantity and quality of water. They are sensitive to the salinity it may contain and to changes in quality, as they adapt to the conditions in which they grow. The quality of the irrigation water affects the nutrition of citrus fruits both because of its content of nutrients in solution and because of the presence of ions that are toxic for the plant.

Among the former, some cations such as $Ca2+$, $Mg2+$ and $K+$ can be found in groundwater and in high concentrations, which can make a significant contribution. Among the toxic elements, the chloride ion stands out, which is generally the cause of salinity. The presence of boron in irrigation water can also cause significant toxicity to citrus fruits (Soler, 2006).

The water requirements of citrus trees, estimated according to their evapotranspiration losses, are between 7,500 and 12,000 m3/ha per year, which is equivalent to an annual rainfall of between 750 and 1,200 mm. But it must be adequately distributed to meet the needs of the crop. None of these factors are present in Mediterranean climatic conditions, so irrigation is essential (Agustí, 2003).

The distribution of irrigation throughout the different months of the year is not uniform, but varies according to transpiration and evaporation, reaching maximum levels in the summer months.

1.3. Pests and diseases

1. 3. 1. Pests

Wonder mite (Aceria sheldoni)

It is a small, elongated, subcylindrical eriophyte. It needs to be protected, so it lives inside the buds in formation, feeding on cell juices (MAPAMA, 2014). Active all year round inside the buds, the periods of greatest activity coincide with spring and summer sprouting. The cycle takes 10-12 days (Lucas, 2009).

As it bites the cells to feed, it causes a series of disturbances which affect the formation of the floral elements, the leaves and the growth of the bud itself. The mites move with the new shoots to the buds in formation. If they set, they end up prematurely dropping the fruit and forming what are known as monsters.

An acceptable threshold for effective pest control has been set at 50% of buds with mites or 25% of branches with deformations. Ideally, most of the shoots no longer than 5 cm in length should be treated (Lucas, 2009). Chemical control is the most effective solution against the pest. To prevent this mite, it is recommended to avoid excessive vigour of the plant, as continuous sprouting can favour proliferation and increase its damage (MAPAMA, 2014).

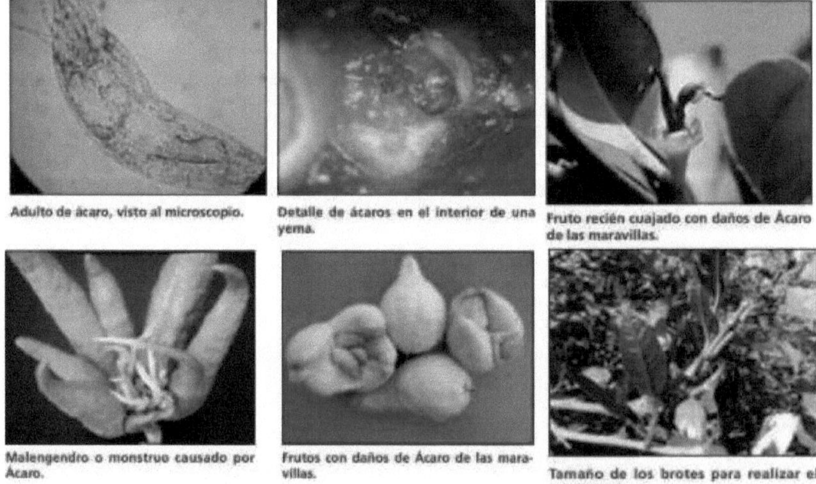

Figure 8: Adult and damage of marigold mite (Lucas, 2009).

Red mite (Panonychus citri).

The adult female is rounded and dark red or purple in colour, with long quetae on the idiosoma, the basal tubercles of which are the same red as the rest of the integument. The male, on the other hand, has a more pearly shape and longer, whitish legs. The eggs are rounded, reddish and with a vertical hair. The remaining stages of development are also red.

Panonychus citri lives preferentially on young, fully grown leaves, but is also found on fruit and green branches. Both adults and immatures are found all over the leaf surface. They prefer to lay their eggs along the midrib on the upper side of the leaf (MAPAMA, 2014).

Its multiple stings produce a diffuse, dull-looking, whitish-white discolouration on leaves, green branches and fruit over their entire surface. The discolouration of the fruit lasts until harvest and causes significant aesthetic damage (MAPAMA, 2014). The critical period for the crop is late summer and early autumn (Lucas, 2009).

It is recommended to treat when the percentage of leaves with phytoseiids is less than 30% and the percentage of leaves with red mite is more than 20% between August and October, and 80% the rest of the year. The most effective natural enemies of the red mite are found among the phytoseiid mites *Amblyseius californicus, Phytoseiulus persimilis Typhlodromus phialatus, with the* species *Euseius stipulatus* standing out (MAPAMA, 2014). If the threshold is exceeded, phytosanitary treatments are recommended. It is recommended to carry out cultural work that allows the elimination of weeds, or pruning, which favours the aeration of the trees (Lucas, 2009).

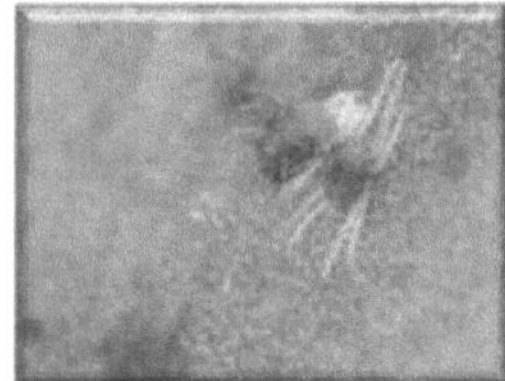 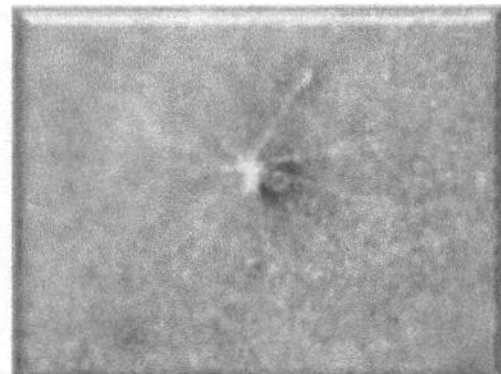 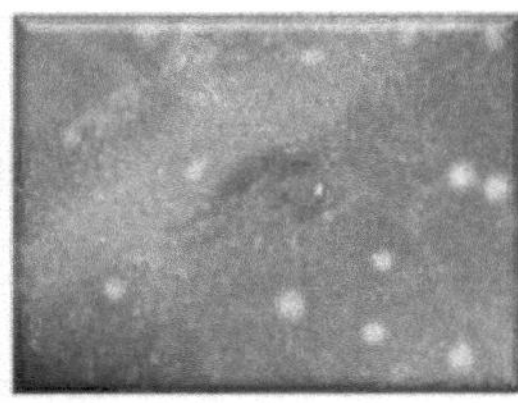

Adult red mite.   Red mite egg.   Euseius stipulatus.

Illustration 9: Adults, egg and natural enemy of the red mite Panonychus citri (IVIA, 2020).

Red spider mite (Tetranychus urticae).
The spider mite lives in colonies on the underside of leaves. They produce large numbers of silk threads, which serve as a refuge from predators and acaricides. In addition, it creates a microclimate which protects it from unfavourable environmental conditions.

It has a very rapid life cycle. After hatching, the mites go through several mobile immature stages. During moulting, it remains immobile and fixed to the substrate from which the adult emerges. The seasonal evolution of population abundance in specific plots is quite irregular, with attacks occurring from May to November (MAPAMA, 2014).

When they feed on leaves they cause discolouration and desiccation, which in most cases is manifested by yellowish spots and/or buckling on the upper surface. They can cause intense and sudden defoliation, especially in summer. They also feed on the fruit, which acquire diffuse rusty spots all over the surface of the ripe fruit.

Treatments are recommended only when the percentage of occupied rings exceeds 54% and the percentage of symptomatic leaves occupied by *T. urticae* exceeds

22% (MAPAMA, 2014). Phytoseiid mites *(Neoseiulus californicus and Phytoseiulus persimilis) and* larvae and adults of the coccinellid coleopteran *Stethorus punctillum* are often seen among their colonies. If the threshold is exceeded, phytosanitary treatments are recommended. Planting fescue between rows increases the level of phytoseiids and reduces the presence of spider mites on the trees (IVIA, 2020).

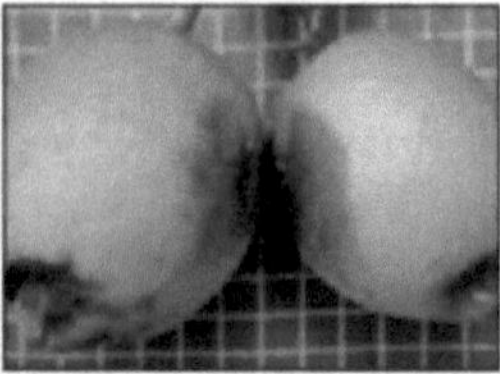

Spider mite adult.

Yellow spider mite damage on citrus leaves.

Moustache or yellow spider mite damage on fruit.

Illustration 10: Adult and damage of red spider mite Tetranychus urticae (Lucas, 2009)

Grooved mealybug (Icerya purchasi).
The most notable morphological characteristic of the adult female is the elongated, ribbed and cottony ovisac. Inside the ovisac are the reddish-orange eggs. The nymphs are reddish in colour with small white waxy secretions on the dorsum and darker legs. During their development, female mealybugs moult three times, passing through three nymphal stages before reaching the ever-motile adult stage (MAPAMA, 2014).

Damage occurs on the plant as a result of uncontrolled mealybug proliferation, damage is confined to the presence of mealybugs, cottony mass and black scales. The critical period for the crop is from spring to October (Lucas, 2009).

The grooved mealybug is controlled by the coccinellid *Rodolia cardinalis.* Chemical control is not recommended. Exceptionally, in very singular cases of very severe attacks and in the absence of *R. cardinalis.* It is recommended to keep the trees well pruned, ensuring proper aeration and avoiding areas with dense vegetation and preventing the ants from climbing the trees (MAPAMA, 2014).

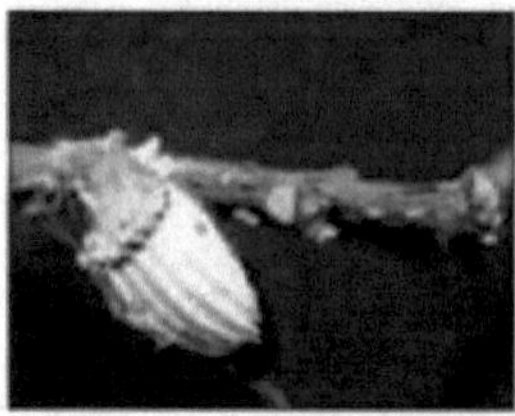

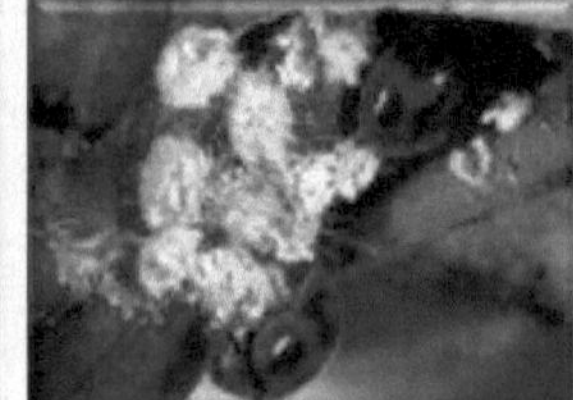

Adulto de cochinilla acanalada con ovisaco y larvas.   Colonia de cochinilla en una rama.   Adulto de Rodolia.

Figure 11: Adults and predator of the grooved mealybug Icerya purchasi (Lucas, 2009).

Cotonet (Planococcus citri)

The pest survives under leaf litter, in cracks or some other protected area of the trunk. When the winter is over, the mealybugs become active and move to colonise the fruit. The critical period for the crop is June (MAPAMA, 2014).

The females are oval and covered with a white waxy secretion and are fertilised by the winged, flying males. They form an ovisac attached to their body containing eggs (between 100 and 200), from which larvae emerge as they hatch and move up the tree (Lucas, 2009). First and second instar female nymphs are oval and pinkish to orange in colour and darken over time. Third instar nymphs are similar to females, but smaller in size (MAPAMA, 2014).

Cotonets cause direct and indirect damage. Direct damage is due to the chlorotic spots they produce on the fruit when they feed on them. This damage is usually observed when fruits are in contact with each other. Indirect damage is due to the secretion of honeydew, from which black mould develops and covers fruit, leaves and branches. In addition, the presence of cotonet attracts other pests such as the citrus borer *Ectomyelois ceratoniae* and *Cryptoblabes gnidiella* (IVIA, 2020).

If the populations of natural enemies are respected and the ants are prevented from climbing the trees, it should not be necessary to intervene. Treat only if more than 20% of the fruit is infested with cotonet (MAPAMA, 2014). Where it causes problems, it is recommended to release the predator *Cryptolaemus montrouzieri* in early spring and the parasitoids *Anagyrus pseudococci and Leptomastix dactylopii* in June. Pruning is recommended to prevent tree branches from touching the ground, removing weeds, locating and placing sponges around the trunk to periodically impregnate them with pyrethrin in order to block the ants' passage to the upper areas of the tree, etc. (IVIA, 2020).

Cotonet adult.

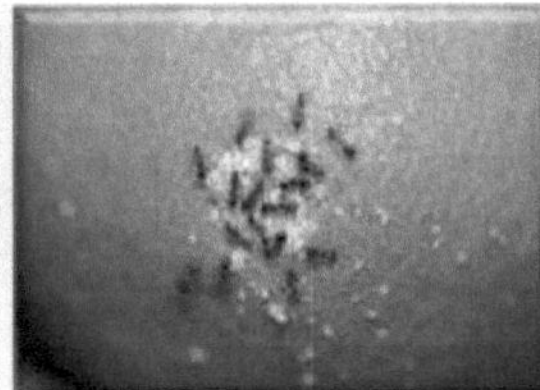

Cotonet nymphs cared for by ants.

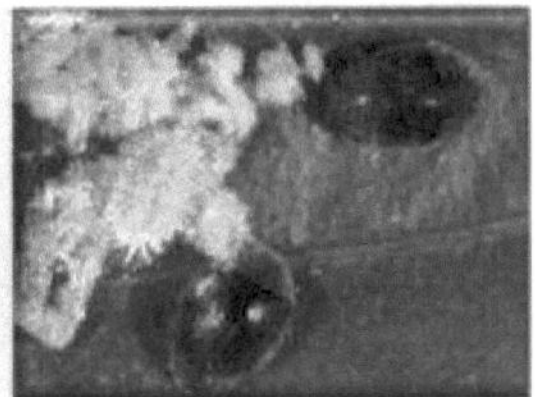

Adults of Cyptolaemus motrouzieri

Illustration 12: Developmental stage of Cotonet and adult predator of Cryptolaemus montrouzieri (IVIA, 2020)

Citrus leaf miner (Phyllocnistis citrella).

It is a micro-lepidopteran that damages tender leaves. The female lays her eggs on tender leaves, from which larvae emerge and penetrate the epidermis of the leaf, making galleries between the dermis. Once the larva has completed the cycle, it emerges from the edge of the leaf and forms a small fold in the leaf, inside which it makes the chrysalis (Lucas, 2009). The adult is about 3 mm and males and females are similar (IVIA, 2020).

Attacked plants decrease their photosynthetic capacity and new plant mass. In adult plants it does not affect production (IVIA, 2020). The greatest development is reached in midsummer, and is always parallel to the presence of shoots susceptible to attack (Lucas, 2009).

Treatments on adult trees in full production are not recommended. It is advisable to treat only grafted, seedlings and developing trees, biological control may not be sufficient and the use of pesticides may be necessary and for this purpose it is recommended to intervene while there are vegetative shoots susceptible to attack and the presence is confirmed (IVIA, 2020). The intervention threshold is 10-30% of infested shoots (MAPAMA, 2014).

It is advisable to encourage biological control by conserving and enhancing the native useful fauna, complemented by parasitoids such as *Citrostichus phyllocnistoides*. The use of protective netting covering the plantations is recommended. In adult plantations, fertilisation, pruning and irrigation can be regulated in order to achieve uniform, short but intense budding, which avoids the continuous presence of shoots on which the leafminer can develop (Lucas, 2009; IVIA, 2020).

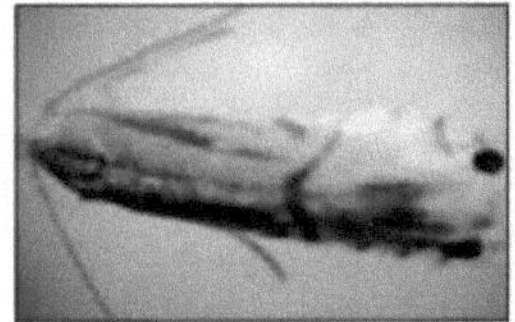

Adulto de minador de las hojas.

Brotes con daños severos de minador de las hojas.

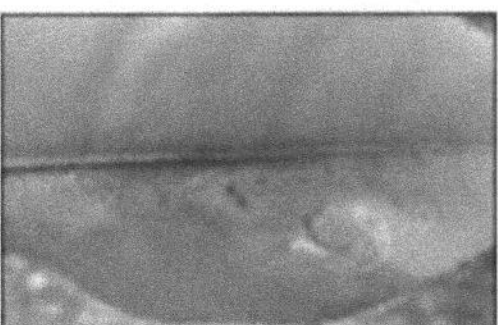

Detalle de galería de minador de las hojas.

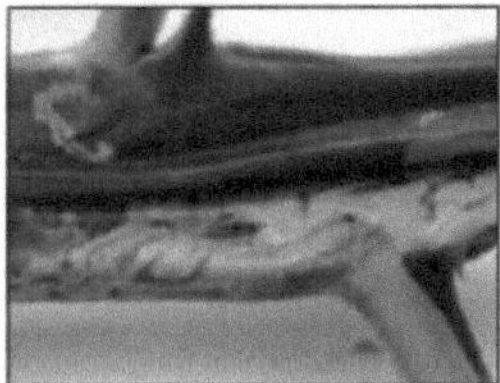

Daños de minador en madera tierna.

Momento idóneo para el tratamiento (brote derecho).

Suelta artificial de parásitos contra minador.

Illustration 13: Adult and damage of leaf miner (Lucas, 2009)

Cottony Fly (Aleurothrixus floccosus) and Whitefly (Paraleyrodes minei)
These pests, being the subject of the present work, are dealt with more extensively in section 1.3.

Fruit fly (Ceratitis capitata Wiedemann)
The fertilised females bite the rind of the fruit and oviposit a few millimetres deep into the rind. After a few days, these eggs hatch and the larvae move into the fruit, feeding on the pulp. The affected fruits end up being infected by pathogens (mainly fungi of the genus *Penicillium),* which causes the infested fruits to fall.

The most recommended chemical method of control is bait treatment, which consists of adding a food attractant to the insecticide used. It is applied to 1 m 2 on the south side of the tree and at a height of 1 metre. Another method increasingly used is mass trapping.

Illustration 14: Adult Ceratitis capitata (University of Florida)
California red louse (Aonidiella aurantii Maskell)
It is considered one of the most damaging pests of citrus fruit, and this is more

21

intense in areas where the fruit is marketed fresh in Spain. This diaspin feeds on the parenchymatous tissue of the plant, where it inserts its stylet into the cells. The symptoms it causes are chlorotic spots on leaves and fruit, due to the destruction of chlorophyll in the mesophyll cells.

In Spain, chemical control is recommended. Good parasitism results have been obtained with ectoparasitoids of the genus *Aphytis,* especially *A. melinus and* A. *lingnanensis* (García-Marí, 2012). Another way of controlling this pest is sexual confusion, and effective systems are currently available for *A. aunrantii.*

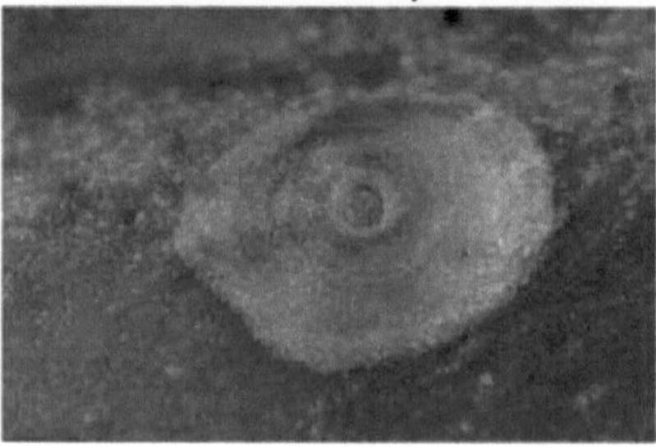

Illustration 15: Adult female A. aurantii (IVIA).

Aphids (Aphis gossypii, Aphis spiraecola, Myzus persicae, Toxoptera aurantii).
They are small sucking insects, preferably found on tender shoots, forming very abundant colonies. Depending on the species and stage of development, they can be of different colours. It is common to find different evolutionary stages in the colony (winged adults, winged adults, winged adults, larvae of different ages) (Lucas, 2009).

The aphids dig their beaks into the tissues and feed on the cell juices, causing defoliation of the attacked shoots, alterations in the growth of the branches and deformation of the leaves. The critical period for the crop is spring-summer for most varieties, coinciding with the beginning of bud break (Lucas, 2009).

Treatments are recommended when 30% of the shoots are attacked. There is an abundance of spontaneous fauna that effectively helps to control aphids. These include lacewings, parasitoids (*Lysiphlebus testaceipes, Aphidius matricariae*) and various coccinellids (*Scymnus*). If the treatment threshold is exceeded, chemical treatment is recommended. Planting of *Festuca arundinacea* as ground cover is recommended (MAPAMA, 2014).

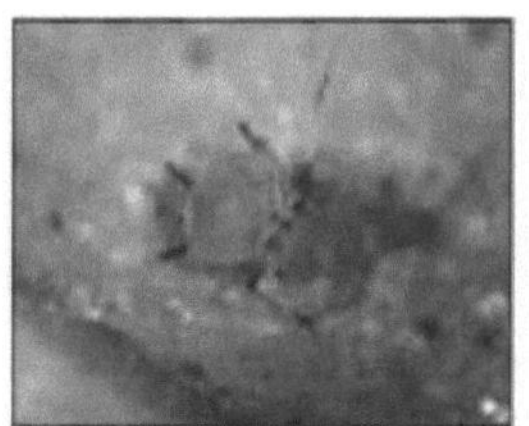

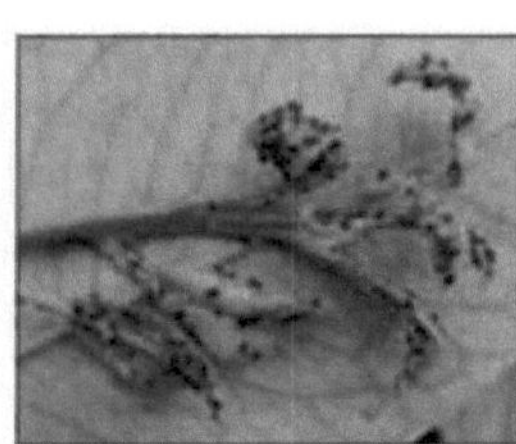

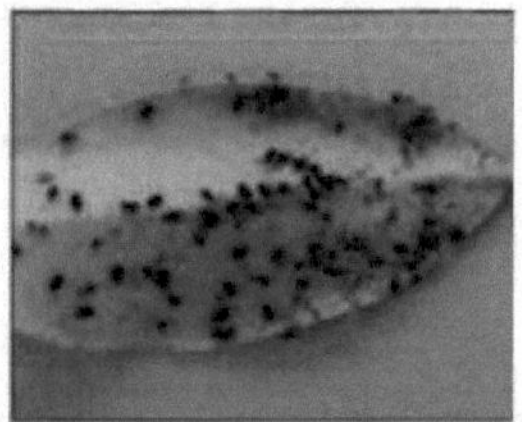

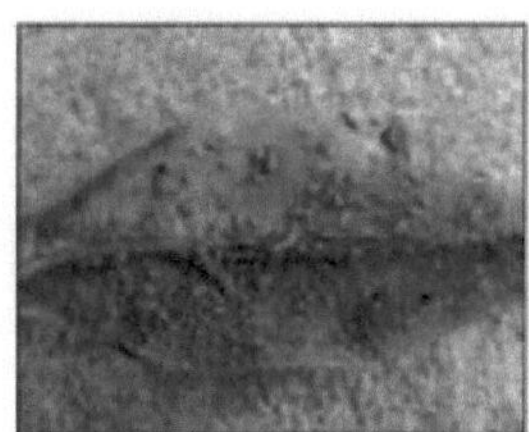

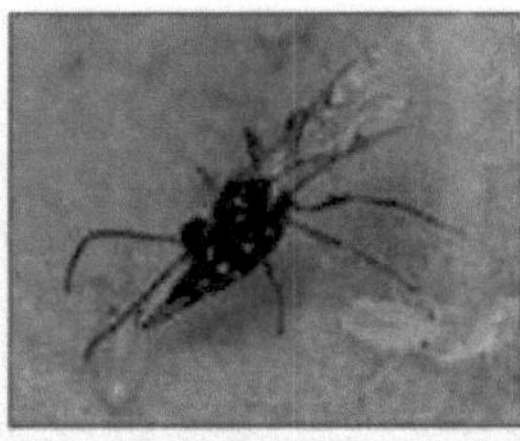

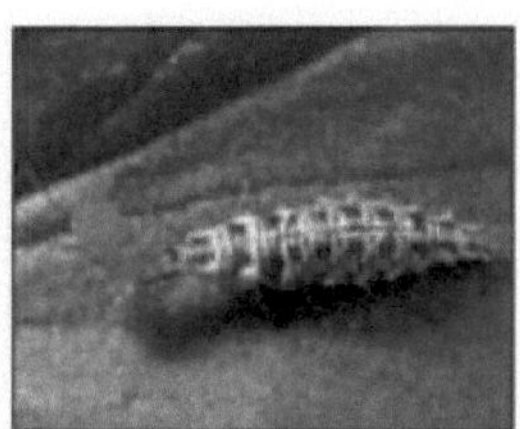

Illustration 16: Aphid adult, damage and natural enemies (Lucas, 2009).

Thrips (Pezothrips kellyanus)

Adults of *P. kellyanus* are brownish-black with pale wing bases, ranging in size from 1 to 2 mm. Females are slightly larger than males and have flared abdomens. They usually appear in aggregates on flowers and young leaves and the critical period for cultivation is from mid-May to the end of June (IVIA, 2020).

Larvae are smaller and whitish, yellowish or orange in colour. The nymphs, which cause the damage, are wingless, white during the first instar and become more yellowish and orange during the second instar (IVIA, 2020).

It causes two types of damage, scarifications or more or less circular scars around the stalk on small fruit, and silvery or discoloured areas on fruit in contact or over the entire surface of the fruit on mature fruit. (MAPAMA, 2014).

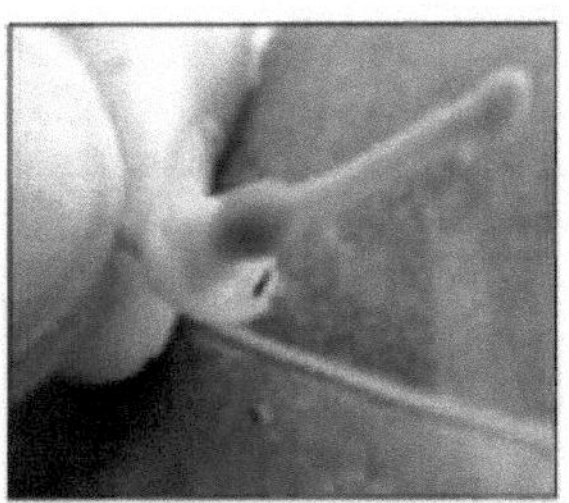

Figure 17: Adults and damage of Pezothips kelluanus on fruits (Lucas, 2009).

1. 3. 2. Diseases

Fruit watering (Phytophthora spp.).
Watery mildew or brown rot disease is caused by several *Phytophthora* species. Waterlogged soil conditions, due to rain or excessive irrigation, favour the development of *Phytophthora* in the plot (IVIA, 2020).

Symptoms are characterised by the appearance of brown soft rots, which progressively advance until the whole fruit is completely affected (IVIA, 2020). In our conditions, the critical period for infections occurs during the autumn months, when the fruit on the tree coincides with heavy rains and mild temperatures (MAPAMA, 2014).

Disease control is mainly preventive, so there is no threshold for action. However, curative action with systemic fungicides is possible, as long as infections are recent and disease symptoms are not yet observed.

Cultural measures should ensure that rain splashes which spread the infective propagules of *Phytophthora* do not reach the fruit in the lower parts of the canopy. It is recommended to prune the skirts of the trees or to raise the lower branches by means of stakes. Maintaining a vegetation cover during the autumn months reduces the impact of rain on the soil surface (MAPAMA, 2014).

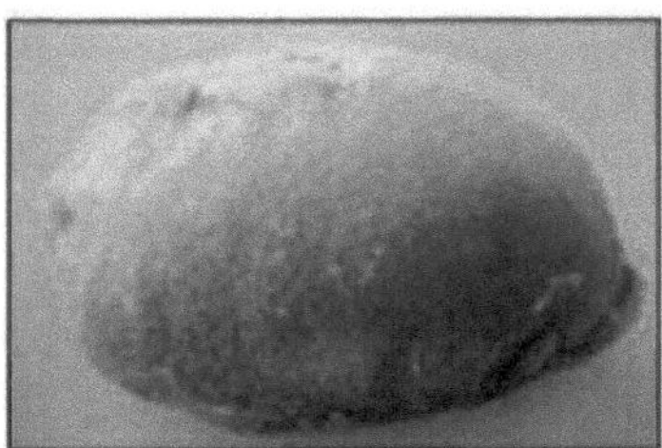

Illustration 18: Lemon fruit affected by Aguado due to Phytophthora (Lucas, 2009)

Neck rot and gummosis (Phytophthora spp.).
This is an internally developing fungus whose presence and activity causes severe

alterations in the sap-conducting vessels and which shows itself to the outside world in the form of cankers and exudates which start at the root collar and in the lower part of the trunk. The affected bark cracks and, when lifted, appears dark yellow underneath. The leaves are affected by the action of the fungus on the vessels, with the nerves turning yellow and only the periphery of the leaves remaining green (Lucas, 2009).

Damage depends on the extent of cankers in the collar and root zone. When the entire periphery is affected, the tree dies. In intermediate situations, progressive weakening occurs, with loss of crop and progressive decrepitude. End branches may dry out under extreme conditions (Lucas, 2009).

Disease control is strictly preventive, so there is no threshold for action. Eradication of established tree infections is difficult and costly. Cultural measures are essential for the control of citrus neck rot and gummosis. In general, situations that favour prolonged waterlogging of the plot should be avoided (MAPAMA, 2014).

To avoid direct water contact with the trunk, it is recommended to grow on plateaus and to keep the drippers separate from the base of the tree. The choice of rootstock is very important, as its susceptibility to *Phytophthora* is highly variable (IVIA, 2020).

Illustration 19: Symptoms of collar rot and gummosis on trees due to attack by Phytophthora spp.

1. 3. Cottony Fly (*Aleurothrixus floccosus*) and Whitefly *(Paraleyrodes minei)*
Plants belonging to the genus *Citrus* are frequently infested by insects of the *Homoptera Aleyrodoidea* group. More than 60 species of whiteflies have been detected on citrus (Mound & Halsey, 1978). Most of these whitefly species usually cause damage in tropical and subtropical areas between latitudes 30° north and 30° south, although occasionally some may cause damage in more northerly latitudes (Mound, 1973). This implies that they do not have a clear winter diapause, but their development in this period slows down and populations decline during cold or dry periods (Gerling, 1990).

Due to the introduction and spread of new species, the colonisation of whiteflies in the Spanish peninsula is similar to that of the rest of the Mediterranean coastal countries. Five species are reported in the Spanish peninsula: *Aleurothrixus floccosus, Dialeurodes citri, Parabemisia myricae, Bemisia hancocki and Paraleyrodes minei* (Garrido, 1991).

1. 3. 1. Taxonomic framework

Order: *Hemiptera*

Order: *Hemiptera*

Suborder: *Homoptera*

Suborder: *Homoptera*

Family: *Aleyrodidae*

Family: *Aleyrodidae*

Genus: Aleurothrixus

Gender: Paraleyrodes

Spice: *Aleurothrixus floccosus*

Species: *Paraleyrodes minei*

1. 3. 2. Geographical distribution
Aleurothrixus floccosus is native to tropical and subtropical America. It is already established in Florida in 1900 and in 1966 in southern California (deBACH, 1975). Its introduction in the Mediterranean basin is relatively recent. In the Iberian Peninsula it was detected for the first time in 1968 in Malaga, and at the end of 1969 it was observed in the province of Alicante, spreading rapidly to all citrus-growing areas, and initially causing considerable damage (Garrido, 1994b).

Paraleyrodes minei laccarino is a species of whitefly described from specimens found in Syria (Laccarino, 1989), although it is a species of neotropical origin whose occurrence was described in California in 1984 and in the Mediterranean basin is relatively recent (Bellows, et al., 1998).

1. 3. 3. Description and biology of *Aleurothrixus floccosus*

	Egg	**N1 (2 phases, mobile and fixed)**
The eggs are light yellow in colour at first, and then evolve to a darker colour as they mature, until they reach a dark brown to black colour, at which point they hatch and the larvae appear.	Pedicel very short, erect, arched, on a patina (substrate, a kind of white lacquer). White colour. Darker when mature.	The larvae are oval and flattened, pale green at first, evolving to yellow in their last stage of development. Transparent, mobile N1 with 8 dorsal serigenous tubercles. Colour caramel , shape

Puesta en semicírculos, incluso círculos completos		alargada. Inicio de excreción de melaza.
N2	N3	N4
Amarillo brillante, pasa de 8 tubérculos dorsales a 6 más desarrollados, así como también aumenta su tamaño. Elevada abundancia de melaza. Secreción cérea perimetral. Más pronunciada en su zona abdominal en forma de flecos.	Ausencia de los tubérculos serígenos dorsales. Secreciones céreas filamentosas marginales que cubren casi todo el cuerpo dorsalmente, pero sobre todo la zona subdorsal o basal.	Última muda. Igual que la N3, pero con el cuerpo oculto bajo las secreciones algodonosas.
Pupa	Adulto	Sustrato/hoja
Ojos compuestos. Idéntica morfología que N4.	El adulto es una mosca de 2-4 mm de envergadura, con cuatro alas recubiertas de un polvillo blanco. Antenas finas y cortas, disposición de alas paralelas, no tienen las alas de forma triangulas.	La plaga se localiza sobre las hojas del cultivo. Hojas muy tiernas, desde cuando apenas hayan alcanzado un 1/3 de su desarrollo, hasta el final de este.

1. 3. 4. Description and biology *of Paraleyrodes minei*

	Egg	N1 (2 phases, mobile and fixed)
Underside of the leaf close to the midrib. Unordered laying, around the nest formed by the female, whose average fecundity is 44.6. Nests built with	Pedicel long, lying, white to cream-coloured, with a yellow spot on the apical part. The development time ranges from 22.2 days to	N1 mobile with perimetral waxy discharge and dorsal cottony discharge. Transparent/reddish in colour. nymphsfixedwith preferenceinplaces

filamentos céreos y que dan el aspecto característico a la puesta de esta especie.

En ocasiones asociada a restos de poblaciones (Bellows, et al., 1998).

12,9 °C hasta 6,3 dias a 26,7°C.

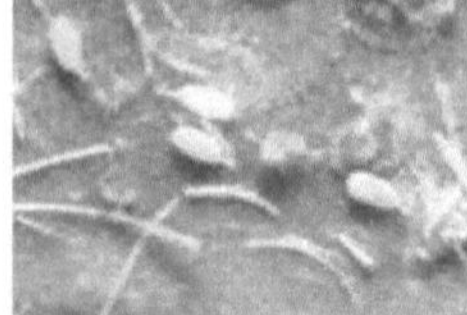

similares a los de los huevos, sin apenas mostrar desplazamientos preferentes y pierde las características morfológicas anteriores. De color amarillo transparente.

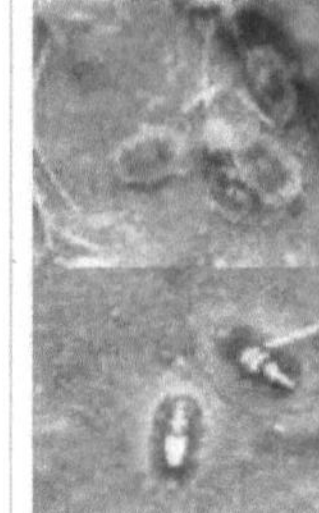

N2

En N2 aparecen unas secreciones céreas rectilíneas longitudinales dorsales. Color amarillo blanquecino, y se caracteriza por ir aumentando el número de secreciones.

Estadio de menor duración ya que se comprende entre 8,6 a 2,1 días.

N3

Secreciones perimetrales, secreciones filamentosas dorsales de 4 glándulas. Lazo de pajarita cercano al orificio anal.

Produce una melaza fina y distribuida por la hoja de forma más homogénea, favoreciendo la aparición de negrilla (García-García, et al., 1992).

N4

Color amarillo, secreciones perimetrales, secreciones dorsales de 10 glándulas, secreciones filamentosas en la zona superior, las cuales llegan a cubrir el cuerpo y 2 glándulas pequeñas en la zona superior central. Lazo pajarita.

El estadio larvario de más duración es el cuarto, que oscila de 15,6 a 5,3 días,

Pupa

Ojos compuestos de color rojizo, hinchado, forma de lazo de pajarita evidente. En la parte posterior del dorso aparecen otros filamentos cereos.

Adulto

Gran tamaño y muy poco móvil, antenas muy largas y gruesas, "mazudas" Alas más triangulares.

Sustrato/hoja

Hojas viejas o maduras. Se asocia a otras especies de mosca blanca.

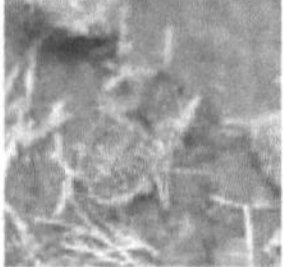

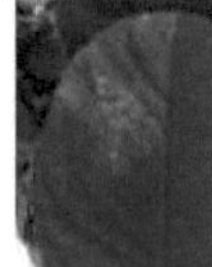

1. 3. 5. Damage

Whiteflies are polyphagous species and feed on plants belonging to more than twenty genera of different families, however, in the Mediterranean area they show an almost exclusive preference for plants of the genus Citrus. (Halsey & Mount, 1978).

Damage occurs to plants when whitefly nymphs penetrate leaf tissues with their stylet to feed. This damage results in reduced crop productivity due, firstly, to the extraction of nutrients from the phloem of the plants. Secondly, and as a consequence of their way of feeding and their digestive system, they excrete honeydew, which can directly contaminate the fruit or cover the leaf surface, causing fungal growth (black mould).

This coverage of the leaves reduces productivity and increases the thermal absorption of the leaf caused by the exposure of dark surfaces to the sun, resulting in a reduction of leaf efficiency. This can cause premature tissue death (Mound, 1973).

It also reduces photosynthesis production, due to reduced light reaching the cytochromes, blocked stomata and limited gas exchange. The presence of honeydew induces the appearance of other pests, such as coccids (Byrne & Bellows, 1991).

The action of the insect on the tree is manifested by a general decay in case of strong attacks, due to the sap suction by the insect. The fruit may also be covered by honeydew, filaments and fungi, making it difficult to process before marketing.

1. 3. 6. Monitoring and Threshold

Monitoring is carried out from spring to the end of summer, because during this period the pest is more aggressive than when the crop is sprouting, which allows it to multiply very quickly and therefore slows down the parasitism process.

For CARM (Autonomous Community of the Region of Murcia) a threshold for *Aleurothrixus floccosus* that could be valid is 10 to 20% of shoots with honeydew nymphs. However, IVIA recommends some kind of action against the pest if the level of infestation exceeds 20% of shoots attacked and the parasitism rate is less than 60%.

1. 4. Integrated Pest Management (IPM)

The Food and Agriculture Organisation of the United Nations (FAO) defines Integrated Pest Management (IPM) as "the careful consideration of all existing pest control techniques and the subsequent integration of appropriate measures to prevent the development of pest populations, keeping the use of pesticides and other techniques to economically justified levels and reducing or minimising risks to human health and the ecosystem. IPC emphasises the development of healthy crops with the least possible interference with the agro-ecosystem, and promotes natural pest control mechanisms".

1. 4. 1. Biological control

Biological control is understood as the action of parasitoids, predators or entomopathogens to maintain the population density of another organism considered a pest at a lower level than would exist in its absence (deBACH, 1975). These pests have natural enemies (parasitoids and predators) capable of total biological control in some cases and partial biological control in others (Soler, et al., 2002).

Most of the predators of whiteflies, especially those that are specific, are small arthropods. Almost all of them are polyphagous, some of them feeding on prey belonging to more than one family. They usually lay their eggs close to their prey populations. They feed preferentially on whitefly eggs and secondarily on young nymphs.

1. 4. 1. 1. 1. Predators

Predators include some mites belonging to the family *Phytoseiidae. Amblyseius spp., Euseius spp. and Typhlodromus spp.* are active predators of whitefly eggs and nymphs as well as feeding on other mites and pollen (Teich, 1966; Osman, 1971; Wysoki and Cohen, 1983; Ragusa et al., 1991; Soto et al., 2016).

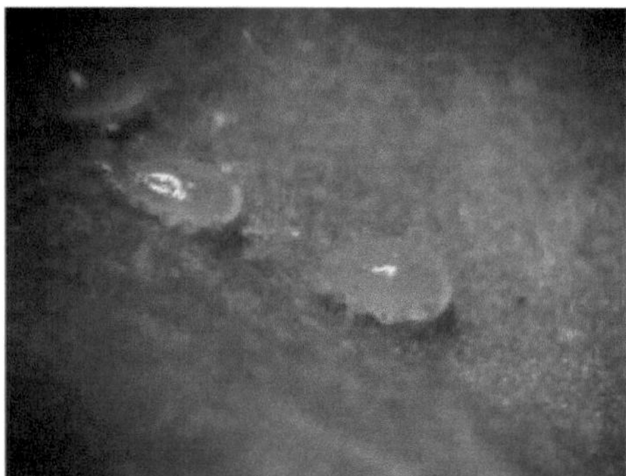

Illustration 20: Phytoseiid mites on orange leaves.

Among the insect predators of this whitefly species are the coccinellids *Clistotethus arcuatus and Cryptolaemus montrouzieri, and the* neuropterans *Chrysoperla carnea and Conwentzia psociformis.* The family of coccinellids includes the most important ones such as *Clitostethus arcuatus* (Rossi), although they are generally not capable alone of keeping whitefly populations below the pest level (Loi, 1978; Bathon and Pietrzik, 1986; Malausa et al., 1988; Gold et al., 1989; Bellows et al., 1992a; Heinz et al., 1994; Hoelmer et al., 1994).

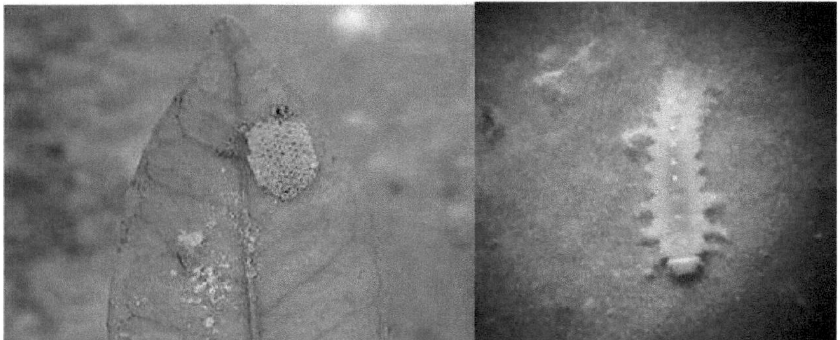

Illustration 21: (Left) Lepidoptera larvae on the upper left and a Chrysoperla carnea larva preying on the colony. (Right) Larva of Clistostethus arcuatus.

1. 4. 1. 2. Parasites

In terms of parasitism, parasitoids create a direct relationship with their host. Parasitoids kill their prey and complete their development outside (ectoparasitoids) or inside (endoparasitoids) the host.

Within biological control, there is a group of entomopathogenic fungi which have the particularity of invading their hosts through the tegument and are therefore considered very useful for controlling the populations of these sucking insects. In fact, there is the possibility of using integrated control programmes of whiteflies, fungi such as *Paecilomyces fumosoroseus, Aschersonia aleyrodis, Verticillium lecanii, Beauveria bassiana and Metarhizium anisopliae* (Albuquerque Maranháo et al., 2009).

Parasites of Aleurothrixus floccosus

Some authors estimate that twelve species of *Aleurothrixus floccosus* parasites are present in seven countries in South America, the area from which this species of whitefly originates. These parasites belong to the genera *Encarsia: Eretmocerus, Amitus, Signiphora and Cales noacki* (deBACH, 1975). *Cales noacki is an* aphelminid native to Brazil which parasitizes 2, 3 and 4 year old nymphs of *Aleurothrixus floccosus*, preferring the 2 year old, which change shape to become more convex and maintain their amber to creamy white colour.

Cales noacki was introduced into Europe in 1970 by France, imported from Chile (Onillon et al., 1972). In Spain, three parasites were introduced in 1970 that had been effective in California and Mexico, *Cales noacki, Eretmocerus paulistus and Amitus spiniferus*, the first one being the one that acclimatised more quickly. The releases were carried out in Malaga, and a year later a notable decrease in *A. floccosus* was observed.

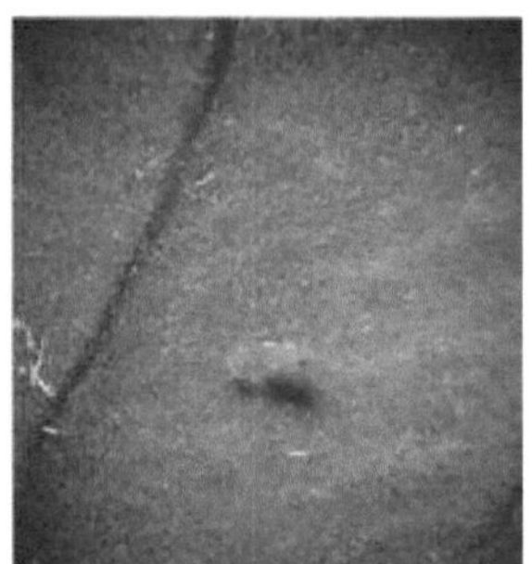

Figure 22: Adult Cales noacki.

The efficacy of *C. noacki* continues to be very good, exerting a good biological control of this species. It can be found in the field throughout the year, but its populations are highest in spring and autumn (Garrido, et al., 1977; Santaballa et al., 1980; Garrido, 1983; Garrido, 1992b). In some areas, *A. floccosus* is also parasitized by *A. spiniferus,* both insects carrying out complementary work (Garrido, 1994b; Garrido, 1994c).

Paraleyrodes minei parasites
No parasites have been found cited on this newly described whitefly species, neither in the publication describing it (Laccarino, 1989) nor in subsequent publications citing its occurrence in various countries (Garcia, et al., 1992; Argov, 1994).

In Spain, natural enemies have hardly been observed in general in the colonies of this species. In contrast, in Israel this species is occasionally found parasitised by *Encarsia hispida* De Santis and in the USA it is known to be parasitised by *Encarsia variegata* Howard (García-Marí, 2018).

1. 4. 2. Chemical control
Royal Decree 1311/2012 of 14 September, which establishes the framework for action to achieve a sustainable use of plant protection products, aims to reduce the risks of the use of plant protection products on human health and the environment, as well as to promote the use of integrated pest management and alternative control methods.

Whiteflies in citrus are very often controlled by the use of insecticides. The high number of annual generations leads to intensified treatments, causing severe damage to the natural enemy complex and facilitating the development of resistance (Dittrich and Ernst, 1990).

For this reason, control measures appropriate to each species are applied, in order to try to optimise resources and respect the auxiliary fauna. Among the measures to be taken into account, the use of authorised products is of great importance, as well as respecting the safety periods and the doses recommended by the manufacturer.

According to the Register of Phytosanitary Products of MAPAMA consulted on 26/05/2020, the active substances authorised for the control of whiteflies in citrus are:

- PARAFFIN OIL 83% [EC] W/V
- ACETAMIPRID 20% [SP] W/W
- Azadirachtin 2,6% (AS AZADIRACTIN A) [EC] W/V
- DELTAMETHRIN 2.5% [EW] W/V
- PYRETHRINS 4.65% (as pyrethrum extract) [EC] W/V
- PYRIDABEN 10% [SC] W/V
- SPIROTETRAMAT 10% [SC] W/V
- SULFOXAFLOR 12% [SC] W/V

With IPC, a balance is achieved by reducing environmental pollution and therefore favouring biodiversity, favouring an increase in auxiliary fauna. On the other hand, the reduced use of phytosanitary products means a reduction in the appearance of resistance. This phenomenon causes a progressive increase in the costs of conventional control, which does not occur when IPM is applied.

1. 4. 3. Cultural measures
- Keep trees well pruned so as to facilitate aeration of the trees.
- Control irrigation and fertilisation to avoid continuous bud break, which would allow more pest attacks on young shoots.
- Fertilise in a balanced way to avoid excessive vigour due to too much nitrogen fertiliser.
- Avoid insecticides harmful to auxiliary fauna such as *Cales noacki and Clistotethus arcuatus.*

2. Justification and objective

Since their appearance in Spain was detected in 1990 (García, et al., 1992), a slow but steady increase in the populations of the whitefly *Paraleyrodes minei* and *Alelothrixus floccos* has been observed in the Valencian Community. This has led to populations becoming established in citrus orchards, causing tree damage and extensive fruit destruction.

For all of the above reasons, the following objectives were set for this Master's thesis:

- Study of the population dynamics of whitefly on Navelina orange by intensive sampling and analysis of the distribution of the population in the different types of leaves on the tree in the summer season.
- Insecticide efficacy trial for the control of whitefly and cottony fly with the approved citrus insecticide Spirotetramat 10% [SC] W/W and Paraffin Oil 83%.
- Analyse the presence of natural enemies and their biological control action on the pest.
- Integrated pest control of *Paraleyrodes minei and Aleurothrixus floccous using* biological, chemical and cultural control techniques.

The aim is to deepen our knowledge of the population dynamics and integrated control of whitefly and cotton fly in view of the damage that has been observed in recent years and the lack of effectiveness and scarcity of resources for their control.

3. Material and methods

3. 1. Location and characteristics of the holding of the plot of land

In order to monitor pests and natural enemies, a plot with a conventional management system of Navelina orange trees managed by Bayer cropscience was used. The Navelina orange trees were planted 8 years ago on a plateau with a planting frame of 7 x 5.5 m, a total of 570 trees, and irrigated by localised drip irrigation.

This plot is located in the Vega Baja del rio Segura region of the Valencian Community (Spain) whose municipality is Albatera. In the illustration 25 an aerial photo of the plot is shown:

Illustration 23: Trial plot boundaries. Albatera.

The plot is divided into 4 sub-plots with the following characteristics:

SUBPARCEL	UTILISATION	CULTIVATION	SURFACE
A	Irrigated fruit trees	Orange	12.264
B	Irrigated fruit trees	Orange	9.738
C	Irrigated almond	Almond	4.928
D	Unproductive	-	

Table 3: Use of sub-plots.

3. 2. Plant material

The trial was carried out on a Navelina plantation which was a mutation of Early Navel originating in California. Vigorous tree (Figure 4), without thorns and dark coloured foliage. Like the rest of the varieties of the Navel group, the fruit has a navel (Figure 6).

Weight	200 - 220 g.	**Juice**	50 - 54%
Diameter	73 - 78mm	**Seeds**	No
Shape Round diameter / height	0,98	**Fructification High**	High
Bark	3,5 - 4,5mm	**Collection**	20 October
Colour Intense orange colour index		**Flower**	Pollen-free

Table 4: Characteristics of the fruit Naranjo var. Navelina.

Early variety, intense orange colour of the fruit, with high levels of acidity maintained until the end of the harvesting period and good adherence to the stalk, although it is quite sensitive to creasing or clearing. It presents some alternation in the harvests. Good tolerance to iron chlorosis and root asphyxia.

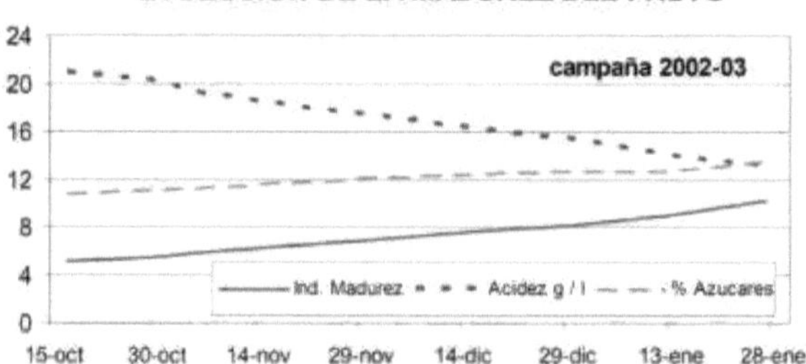

Graph 3: Evaluation of maturity of the fruit Naranjo var. Navelina.

3. Phytophagous sampling

3. 3. 1. Phytophagous sampling in the field

For monitoring, a weekly sampling period was established from 5 July 2019 to 31 July 2019. Due to the treatment on 18/07/2019, sampling was changed to every two weeks from 31 July to 25 September 2019.

The sampling procedure consisted of choosing 40 leaves with *Alelurotrixus floccosus and* 40 leaves with *Paraleyrodes minei* from 20 trees at random and from three strata of the plant (upper, middle and lower), taking each stratum as one third of the height of each plant, resulting in a total of 4 leaves per plant and 80 leaves in total.

Samples of the phytophagous plants were taken with scissors, easily identified in the field with a hand magnifying glass and kept in white plastic bags labelled with the name of the pest. Once the leaves were in their respective bags, they were transferred for identification and counting into a portable cooler with ice bars to keep them in suitable conditions.

3. 3. 2. Laboratory sampling of phytophages

The following literature was used for the identification and differentiation of phytophagous stages:

- Geographical distribution and seasonal and interannual evolution of the whitefly *Paraleyrodes minei* (Hemiptera: Aleyrodidae) in citrus crops in the east of the Iberian Peninsula by Ferra García Marí (2009)
- Biology and seasonal dynamics of *Paraleyrodes minei* by Eduardo Langa Llueca (2018). Polytechnic University of Valencia.
- Bayer's Murcia Food Chain Partnership group's own elaborations 2019.

The pest stages and their severity were identified and counted using a template (Annex 7.4.) developed by Bayer's Food Chain Partnership. In this template, the presence of the stage (Egg, N1, N2, N3, N4, Pupa and Adult) was marked with an "x", and all stages could be present on the same leaf. To describe the population dynamics and structure, the number of individuals found on the leaves and the different immature stages to which the individuals belonged were counted in the laboratory.

Severity is measured on a scale of 1 to 4, the values of which are shown in the table:

Number	Leaf occupancy (%)
1	<25%
	25-50%
	50-75%
	75-100%

Table 5: Leaf severity classification.

For identification in the laboratory, a LEICA MZ6 binocular loupe was used, with a 10x eyepiece, 1.5x objective, and zoom up to 4.0x. Photographs taken with the loupe were taken with a Xiomi Red Mi 4X mobile camera.

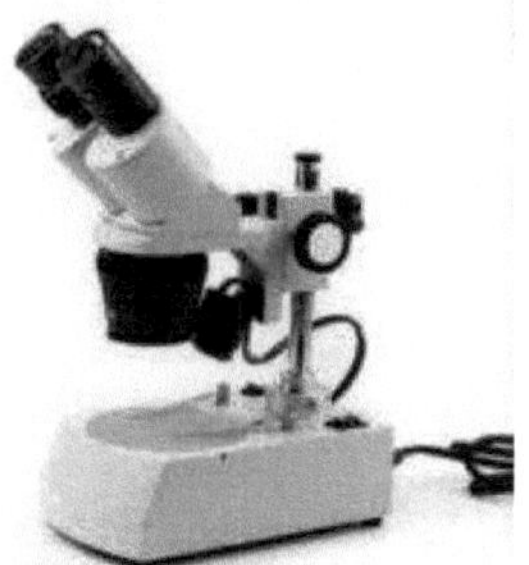

Illustration 24: Binocular LEICA MZ6.

3. 4. Sampling of natural enemies

3. 4. 1. Field sampling of natural enemies

Yellow Econex sticky traps were used to capture useful auxiliary fauna. The chromatic trap has a size of 40x25 cm and is divided into two parts of 20x 25 cm each. Each trap is annotated:

- Date of placement
- Date of collection
- Municipality
- Cultivation
- Variety

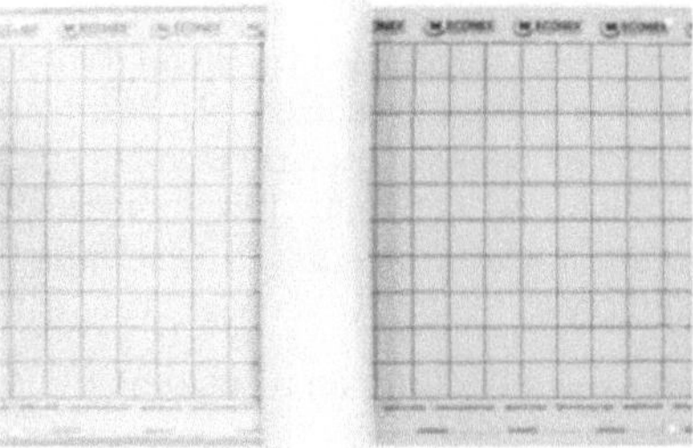

Illustration 25: Econex colour trap.

They were hung in a tree at a height of about 1.5 m, facing south. Once in place, the protective paper was removed from the side of the grids. On collection, the traps were covered with the same protective paper that came as standard and the necessary information was noted down.

Traps started to be placed on 24 July 2019 and were changed every two weeks until 25 September 2019. For this monitoring, only one trap was placed in each plot (Figure 26).

Illustration 26: Positioning of the chromatic trap in the plot. Albatera.

The yellow colour plates were taken to the laboratory for identification and recording of the number of individuals trapped. In this way the population density of natural enemies was compared with the whitefly populations.

3. 4. 2. Laboratory sampling of natural enemies

Each trap was observed through a LEICA MZ6 binocular, quantifying parasitoids and predators and classifying them by family and species in a Bayer Food Chain Partnership template (Annexes 7.2 and 7.3). In the template, the citrus variety, date of placement and collection, type of trap, number of days in place and number of traps were recorded. As for the phytophagous photographs, a Xiomi Red Mi 4X mobile camera was used.

The auxiliary fauna was identified using keys provided in the Course on Sampling, Identification and Management of Pests and their Natural Enemies in Citrus and Persimmon of the Polytechnic University of Valencia and Field Guide: Citrus Pests and their Natural Enemies.

3. 7. Meteorological data

At the same time, meteorological data was taken from the weather station in La Murada (Orihuela), supplied by the SIAR Network (Agro-climatic information system for irrigation) in order to be able to glimpse the influence of climatic conditions on the development of pests and auxiliary fauna.

Province:	Alicante
Term:	Orihuela
UTMX:	678796.000
UTMY:	4227878.000
Spindle:	30
Height:	98m
Date of installation:	09/04/2015

Table 6: Climatic station located in La Murada (Orihuela). IVIA.

3. 8. Application of phytosanitary treatment

In this work, the Bayer Cropscience threshold was used, which recommends a treatment for whitefly and cottony fly control on 60% of leaves occupied by eggs plus first instar nymphs and a severity of 3. In this way, the greatest number of vulnerable instars is used, avoiding the refuge of the pest in waxy secretions and honeydew.

A recommended treatment was carried out on Thursday 18 July 2019 as the environmental conditions were suitable for good efficiency, to avoid fruit spotting or drift problems. The justifications for the measure are:

- Exceeding the treatment threshold for *Paraleyrodes minei* and *Aleurothrixus floccosus*
    - Reducing Cotonet *(Planococcus citri)* populations
    - 2nd generation California red louse (Aonidiella aurantii)
    - Fall in the presence of auxiliary fauna

Phytosanitary application was made with the following active substances and conditions:

Localidad	Cultivo	Ingrediente activo	Dia de la aplicación	Nombre comercial	Dosis/ 1000 lts	Modo de aplicación	Vol de Caldo
Albatera	Naranja	Spirotetramat 10%	18/07/2019	Movento Gold	0,75L	Atomizador	2000L
Albatera	Naranja	Aceite de parafina 83%	18/07/2019	-	1,5L	Atomizador	2000L

Table 7: Phytosanitary treatment

Station: Orihuela - La Murada					Date: 18/07/2019			
T med	T ma x	T min	Average wind speed	Peak gust	Peak gust time	Radiation	Time s sun	Precipitation
26,09	33, 1	20, 01	3.87km/h	14.89 km/h	12:16	27.56 MJ/m2	12,6	0 mm

Table 8: Weather conditions on 18/07/2019. La Murada-Orihuela. IVIA.

Movento Gold (registration no. ES-00024) shows systemic behaviour (upward and downward), acting especially by ingestion. This product is active on a wide group of pests of the homopteran group. It is effective on the first immature stages of this group. This product has a safety period of 14 days and an application rate of 0.4 - 0.75% for citrus.

Illustration 27: 3L bottle of Movento Gold. Agricoal Prime 2020. Paraffin oil 83% w/v. EC is an emulsifiable concentrate without safety pest control (PS: NP), with contact insecticidal activity. It is used to control citrus pests such as mealybugs and whitefly at 1-1.5 L/hl. Treatments are given before fruit colour change.

3. 9. Data Analysis

With the data obtained, pivot tables were created with the Excel tool to classify and order the information. In this way, the population dynamics and structure of *Paraleyrodes minei* and *Aleurothrixus floccossus* were explained. The same application was used to compare the behaviour of natural enemies and the phytosanitary treatment according to the integrated pest management carried out on the plot.

4. Results and Discussion

4. 1 Pest assessment

4. 1. 1. Background to farm management

The plot receives all the usual cultivation practices, including phytosanitary treatments for pest and disease control as shown in the farm history.

Prior to 2017, there were never any whitefly problems. Treated only with Movento and Envidor in 2nd generation of California red louse, coinciding with 2nd generation of Cotonet.

In the spring of 2017, a surplus of Acetamiprid was treated for aphids. Throughout the summer and autumn there were heavy attacks of both whitefly species, so that the fruit had to be cut in December, with significant problems of detritus.

In the 2018 season, the whiteflies were not treated in spring and were monitored at the end of June. *Paraleyrodes minei* caused damage in the first generations, so that it stained the fruit that was most difficult to treat, i.e. the inside of the tree and the upper parts of the tree. In this situation, Movento Gold plus oil was treated at the beginning of July with good efficacy. In August, another treatment of orange oil plus Mancozeb was applied to try to wash the stained fruit.

In September, as the new summer leaves emerged, the adults of *Aleurothrixuss floccosuss* appeared. Weeks later the symptoms of parasitism on *Cales noaki* were present. Completely black pupae, supposedly parasitised by *Amitus spinifferus,* were also observed. A release of lacewings (10000 pcs/ha) was also made, but it was difficult to adapt and was therefore not effective, although many lacewing clutches were subsequently seen near leaves with whitefly.

Aphids were also not treated in 2019. In the monitoring of the whitefly species, it is observed that the first generations have not stained the fruit, due to the predator and parasitoid pressure that has been developed by the integrated control strategy.

4. 1. 2. Population dynamics: *Paraleyrodes minei*

In the introduction, the heat integration of the whitefly is said to be 378oC/day, i.e. the total generation time ranges from more than two months to less than 20 days, which can result in 6-7 generations/year.

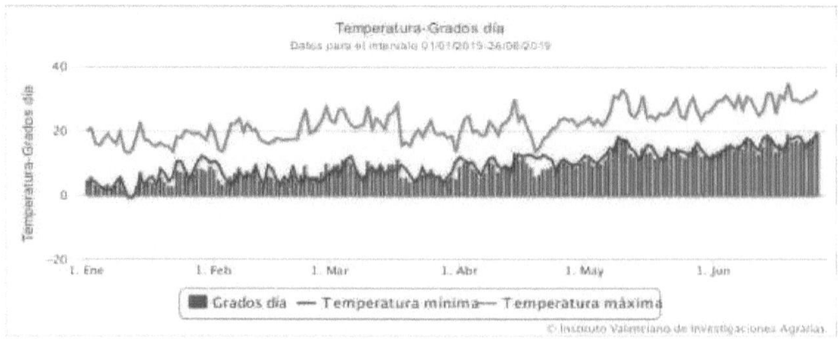

Graph 4: Temperature-Grade days (01/01/2019-26/06/2019)

The average threshold temperature is 7.5oC (the different stages range from 5.7 oC in L4 to 9.2 oC in L2) (Garcia-Marí, 2018), which explains why *Paraleyrodes minei* continues to evolve during the winter months in the Mediterranean citrus-growing area. The presence of adults during this period continues to produce eggs, albeit in smaller numbers. This is a striking and characteristic aspect of this species, as this behaviour is not usually observed in other whitefly species (Soto Sánchez, 2018).

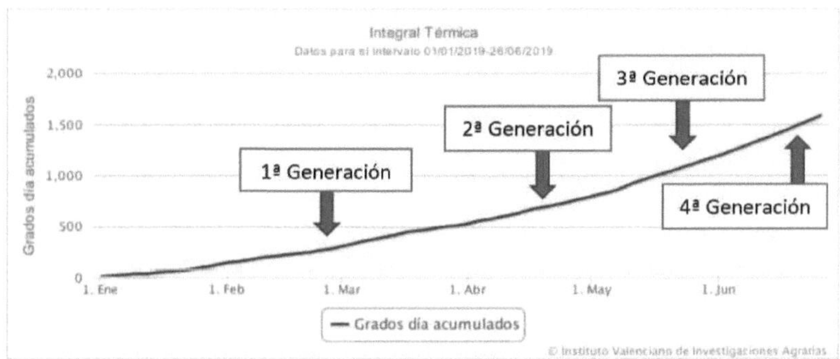

Graph 5: Thermal Integral of Paraleyrodes minei (01/01/2019-26/06/2019)

The months of March and April resulted in very low populations. At the beginning of May, adult flight and oviposition started again. The data observed in the climatic graphs show marked generations of *Paraleyrodes minei* being in the 4th generation at the time of the beginning of the sampling.

As can be seen in Graphs 7 and 8, in June and July the severity is high (>60% of leaf occupied), with an average of 3-4 leaves/branch present. The population increases week by week, increasing the number of eggs and nymphs.

Illustration 28: Orange tree branches with a high presence of Paraleyrodes mieni.

In July, eggs were present on all leaves collected for sampling of *Paraleyrodes minei.* The number of eggs was found to be higher than the number of individuals of the following immature stages.

N1 nymphs are the second most dense stage occupying the leaf. The considerable decrease in the number of immature individuals of the first instar relative to the number of eggs from which they emerged indicated a high level of mortality and predation at this time. In other citrus whitefly species, high mortality occurs at the passage from egg to fixed N1 stage (Soto, 1999).

In the same way, it is seen that this did not happen in the passage to the next instars, since mortality in these instars decreases and is very low. This is typical of other whitefly species in citrus (Llorens, 1994).

Therefore, the stages progressed to more mature stages with a higher pressure on leaves despite the high degree of predation by the EENN. As a constant, generations overlapped and resulted in a large presence of adults reaffixing nests with eggs. New generations were produced, the stages of which are fairly evenly distributed. This implies a difficulty in determining the period of application of the treatments. Other authors have observed development patterns of this species very similar to those shown (García-Marí, 2018).

Illustration 29: (left) Colony of Paraleyrodes minei next to a pupa of Conwentzia psociformis. (Right) Paraleyrodes minei adults laying eggs.

In July 2018, there was a clear problem of spotted fruit, especially in the centres and at the top of the tree. In 2019, the presence of whitefly was very low due to abundant predators, but did not result in fruit spotting and loss of quality.

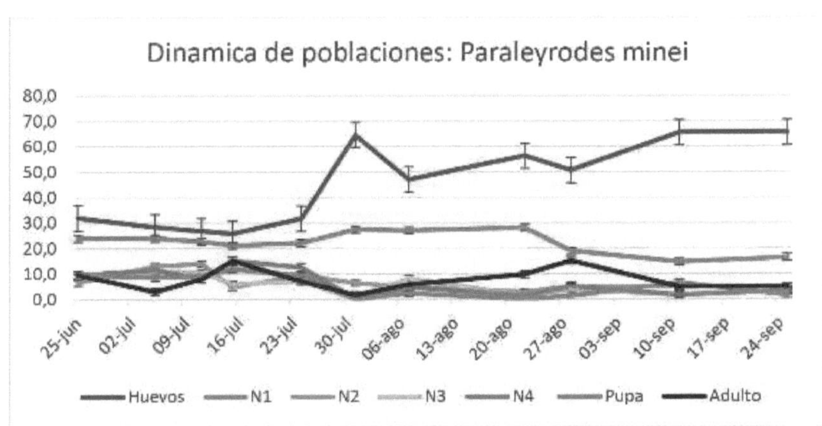

Figure 6: P.minei population dynamics

The population dynamics was discerned to be at its highest in the third week of July, when the sum of eggs and N1 nymphs was higher than 60% of the total and leaf occupancy was higher than 50%.

Knowing well the behaviour and population structure of *Paraleyrodes minei,* control in citrus was carried out by means of pesticide treatments in view of the high population, exceeding the established treatment threshold. The data obtained on the average severity of *Paraleyrodes minei* was modified by the treatment is shown in the graph:

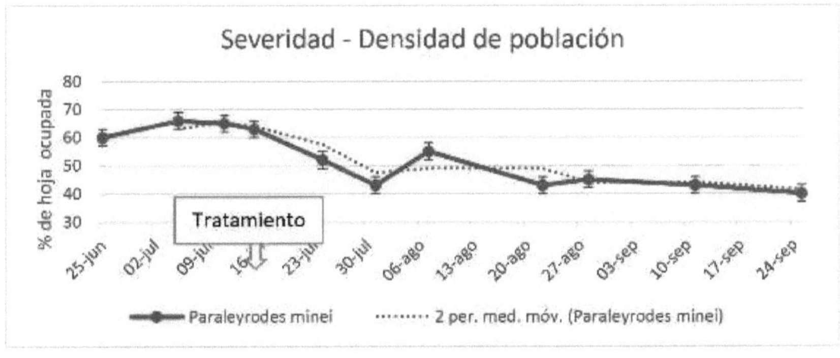

Figure 7: Severity of Paraleyrodes minei

Paraleyrodes minei pressure slowed down from 4-5 leaves/branch on average to 1 leaf/branch. Likewise, the upper severity of 60% of the leaves fell below 50%. Looking at the dynamics, graph 7 shows that the population remained stable until the end of the application.

There were two types of leaves: The first leaves, the outer leaves of the tree, were clean with some eggs and adults laying new eggs. The second leaves, the inner leaves of the tree, showed several dying stages covering more than 25% of the leaf because the treatment did not reach them properly. This results in a reservoir of the pest and may generate a future generation.

Illustration 30: (Left) Branch on the outside of the tree with only 1 leaf present. (Right) Branch inside the tree with several leaves with Paraleyrodes minei present.

Fewer adults were observed nesting and the different nymphal stages were scarce and dead or moribund, i.e. dry and dark-coloured nymphal exoskeleton. In the sampling, all the occupied leaves presented eggs, so the percentage is higher (65.4%) on 30/07/2019 compared to the others.

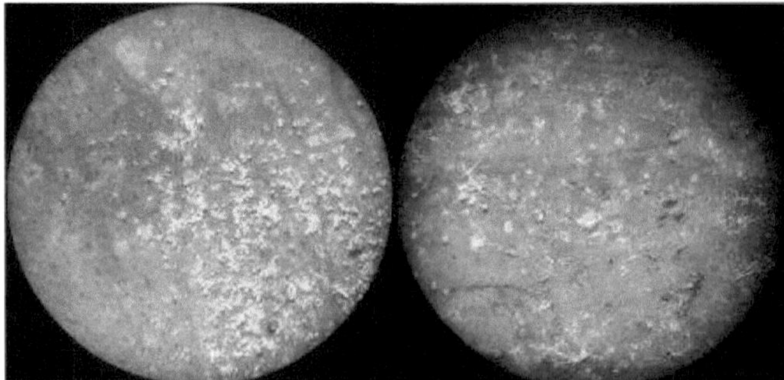

Illustration 31: Leaves with colonies of Paraleyrodes minei after phytosanitary treatment.

In terms of the number of individuals per sample, the presence of eggs predominated in the total number of elements sampled. These eggs were present before the treatment, as the nymphs were eliminated, leaving the previous generations resolved.

There was a greater presence of N1 throughout August. This was due to the large number of eggs that resulted in numerous N1, low predation on *Paraleyrodes minei and* high temperatures. In both field and laboratory sampling, the number of adults and nymphs remained low, with no more than 2 live individuals per leaf.

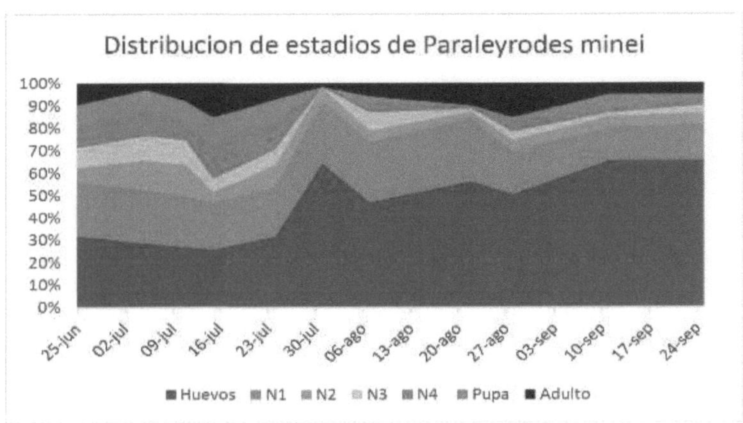

Figure 8: Distribution of stages of Paraleyrodes minei.

In the sample taken on 22/08/2019, a new generation was identified by observing the first leaves with adults carrying out postures and with a large presence of eggs that begin to develop giving the first N1 stages with little predation. In this new generation different stages were observed, but in much less quantity and severity.

Figure 32: Adults of Paraleyrodes minei laying new eggs.

In the last sampling in September, sampling for *Paraleyrodes minei* was difficult due to low population and leaf occupancy (less than 25%). Egg values were lower due to the higher attack on the auxiliary fauna generated in the plot. Approximately half of the leaves collected for sampling were found with egg predation and death of nymphs, i.e. no live individuals.

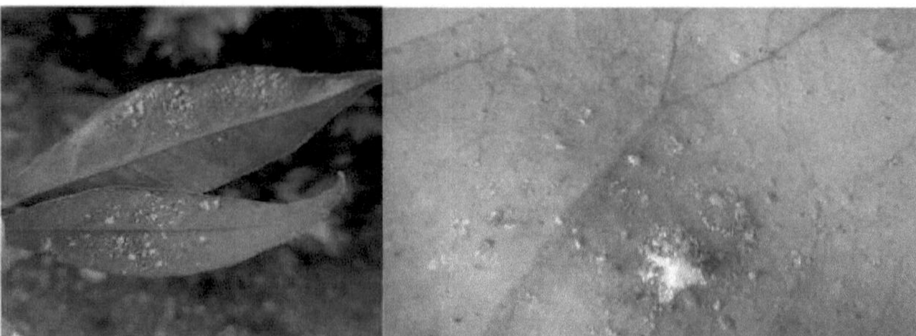
Illustration 33: Colonies of Paraleyrodes minei disordered by predation.

Whiteflies at the end of monitoring were found to be attempting to re-generate with adults laying new clutches. This was observed in the field sampling, but is not reflected in the laboratory, with low severity of adults, nymphs and eggs. On the contrary, the predators are again present in the field making a correct biological control.

4. 1 .3. Population dynamics of *Aleurothrixus floccosus*

Due to the lack of knowledge of the thermal integral of *Aleurothrixus floccosus*, the generation in which it was found is unknown. In winter, due to low temperatures, the intensity of the pest's attack decreases and its populations are also almost nil in March and April (Luppichini et al.,. 2007).

By observation it was believed that 2 generations have occurred and sampling starts in the 3rd generation. Sampling was intended to determine the maximum and minimum population to determine the optimum time to apply a phytosanitary treatment.

The cottony whitefly is a polyvoltine phytophagous pest whose infestations start during active tree growth or in suckers, as the adults lay eggs only on the underside of tender leaves. Therefore, the farm manager removes suckers and excessive vegetative shoots as a cultural control technique.

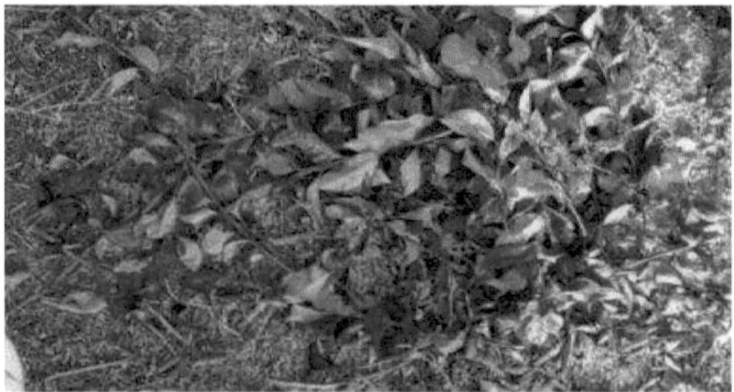
Illustration 34: Removal of orange tree suckers.

As with *Paraleyrodes minei,* there was an overlap of generations. The population of *Aleurothrixus floccosus* was low compared to *Paraleyrodes minei,* occupying no more than 25% of the leaves. In contrast, colonies with abundant whitish woolly and honeydew covering the nymphs were observed, which attracted other insects such as ants, bees, flies and wasps.

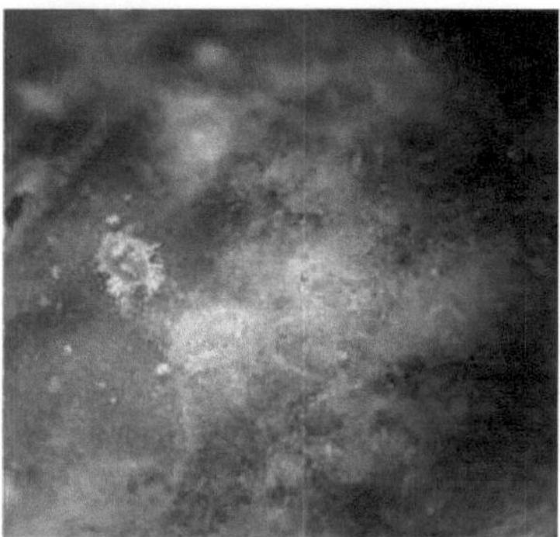

Figure 35: Different stages of Aleurothrixus floccous with abundant wooliness.

The presence of eggs remained constant in July with high predation. This was due to the observation of broken brown eggs (Figure 36. Right) in a disorderly manner and not black in a circular shape characteristic of *Aleurothrixus floccosus.* The passage from egg to N1 nymph had a high mortality, as did *Paraleyrodes minei,* so a low percentage of N1 to eggs was observed.

Eggs, as well as honeydew secretion by the nymphs, were scarce due to high predation and parasitism by *Cales noacki.* This could be seen in disorganised colonies with little or no honeydew.

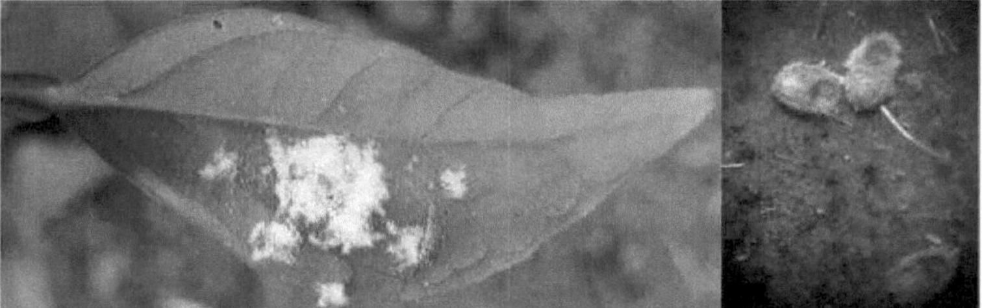

Figure 36: (Left) Conwentzia psociformis preying on a colony of Aleurothrixus floccosus. (Right) Eggs parasitised by Cales noaki.

Mature N3, N4 nymphs and pupae were difficult to find alive as a large presence of exoskeletons with a kite-shaped opening due to *Cales noacki* was observed.

Therefore, N2, N3, N4 nymphs and pupae had no negative impact on the fruit and the tree, i.e. it was below the treatment threshold. Adults tended to fly away so very few were observed in the field and in the laboratory.

Illustration 37: Whitefly and cottony fly on orange branch.

In general, in the absence of insecticide applications, parasitism levels would keep the pest in ranges that do not cause economic damage, but high temperatures reduced the population of *Cales noacki* and other predators. *Alelothrixus flocossus* is therefore under control, although a new generation is dazzling.

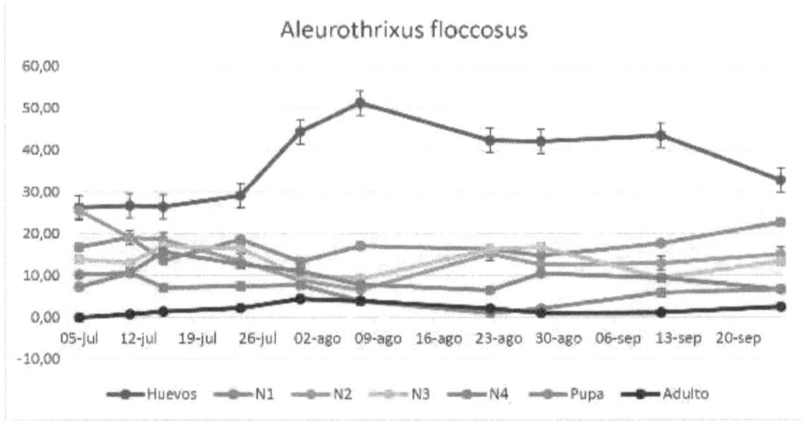

Figure 9: Population dynamics of Aleurothrixus floccosus.

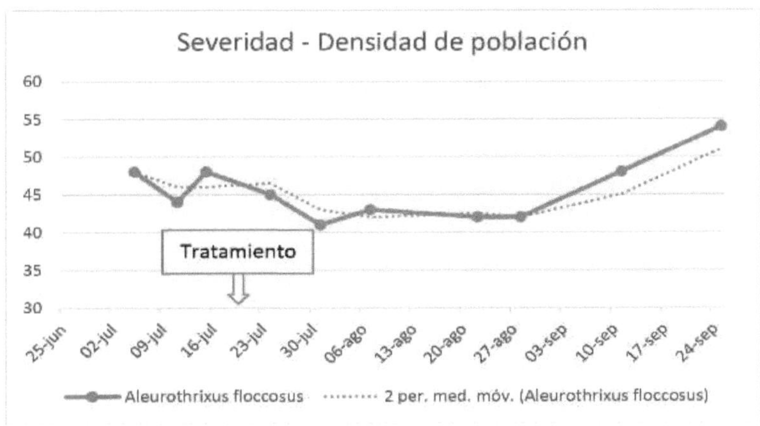

As with *Paraleyrodes minei,* after treatment, a washing of the tree was observed, where the outer leaves are cleaner than the inner leaves. Using cultural techniques, suckers continued to be removed from inside the tree. The treatment had an insecticidal effect, affecting both cottony whitefly and auxiliary fauna.

On the young inner leaves, adults were observed to have started flights for new egg-laying. A clear attempt at generation was identified by the appearance of new colonies with clutches on the characteristic white patina on new leaves. On this patina, some adults were identified in a continuous form with incipient white egg-laying.

Illustration 38: Young leaves inside the tree with Aleurothrixus floccosus

The leaves were mainly egg-bearing as this was related to the new laying that will be taking place. There was also a decrease in the presence of immature individuals related to the treatment.

Figure 39: Adults of Aleurothrixus floccosus laying on white patina.

The new generation started to develop as eggs and the first N1 nymphs were observed darkening near nests, but were little parasitised and predated by auxiliary fauna.

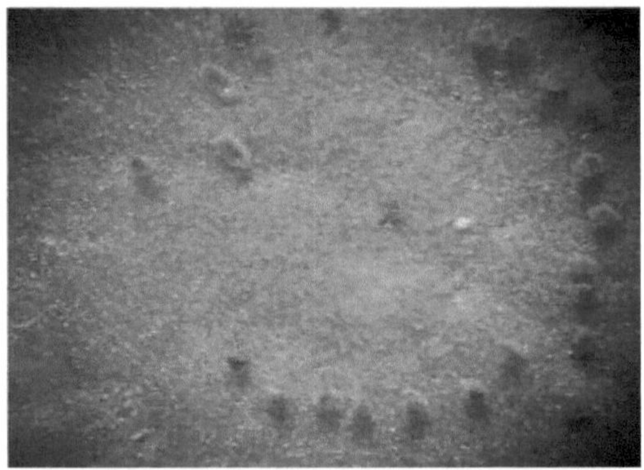
Figure 40: Aleurothrixus floccosus eggs maturing brown to black.

These *Aleurothrixus floccosus* eggs continued to develop, with numerous individuals reaching more mature stages and along with them the secretion of honeydew. The 1st generation of cottony fly after treatment resulted in few whitefly colonies and low honeydew secretion, despite low population parasitism and predation.

Illustration 41: Aleurothrixus floccosus colony without molasses.

At the end of August, coinciding with the decrease of the insecticidal action of the pesticide, *Alelothrixus flocossus* started to lay more frequently and the new generation developed. It gave rise to numerous mature eggs and the first N1 nymphs.

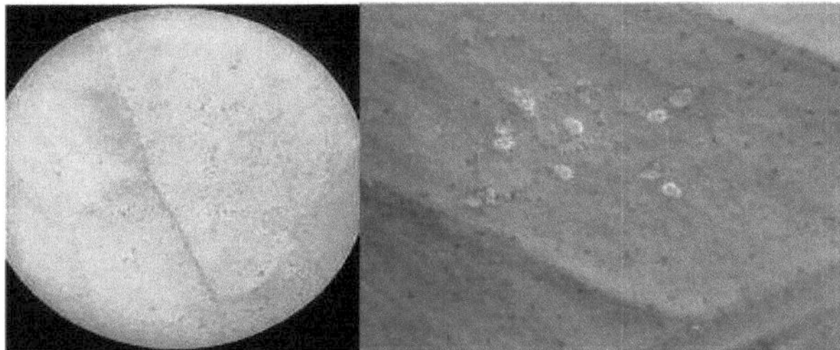

Figure 42: Eggs and N1 nymphs of Aleurothrixu floccosus.

Secretion of honeydew by the numerous N2, N3 and N4 nymphs was more visible, but increased parasitism and predation reduced their incidence. Severity began to increase and several N4 nymphs and live pupae were identified sheltering on the cottony secretions.

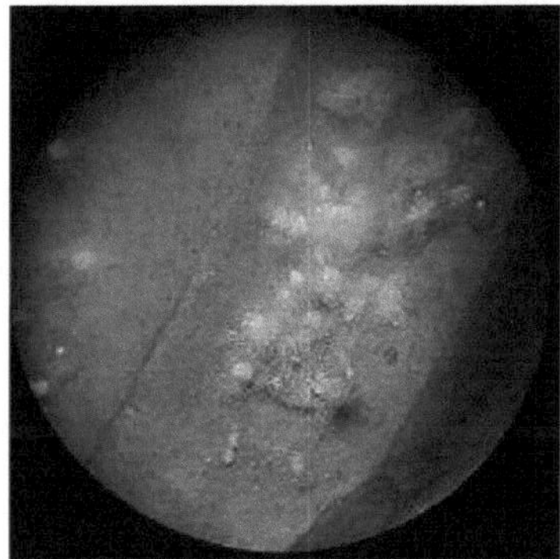

Illustration 43: Colony of Aleurothrixus floccosus forming blotches and honeydew on the leaf.

In September, fewer new clutches were observed because the number of leaves with eggs did not increase. In the field, adult leaves were examined, laying less frequently, but this was not reflected in the laboratory, as the adults flew away.

The presence of cottony fly continued to increase with honeydew secretions reaching 25% of the leaf surface. Cottony fly presence was low, continuing at the 1-2 leaf/branch point, but with an increase in severity above 50%. *Aleurothriscus floccosus was* not able to develop nymphs beyond the N3 stage due to parasitism by *Cales noacki* which started to recover its population.

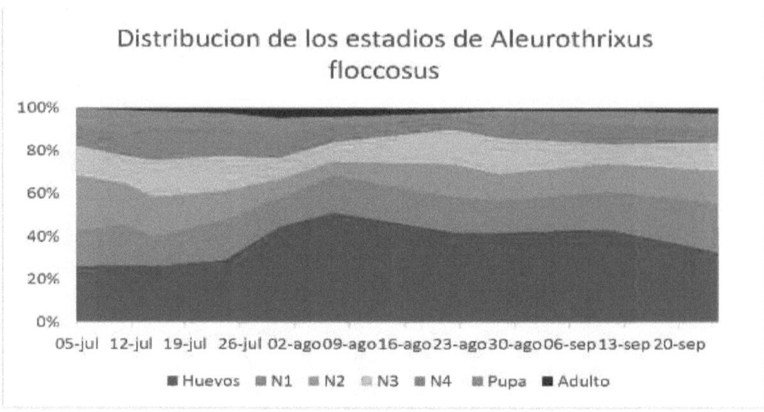

For all these reasons, the generation of *Aleurothrixus floccosus* that developed after treatment has been controlled because of its low incidence. However, the new generation of cottony fly is being more severe partly because of the decrease in natural enemies and the high number of eggs in the clutches.

4. 2. Monitoring of the NSSEs of the sampled period

It is worth noting the variety of species, both parasitism and predators, with the whitefly being the centre of an ecosystem, providing evidence that effective integrated control is being carried out. It is a living ecosystem with predators and parasitoids with food at their disposal.

Predators of highly polyphagous species such as larvae of *Clistostethus arcuatus,* Chrysopas *(Crhysoperla carnea) and Semidasalis aleyrodiformis, as well as* numerous pupae and adults of *Conwentzia psociformis* were observed near whitefly nests. Also mite species *(Euseius spp.)* on leaves inside the tree and Lepidoptera larvae in the laboratory.

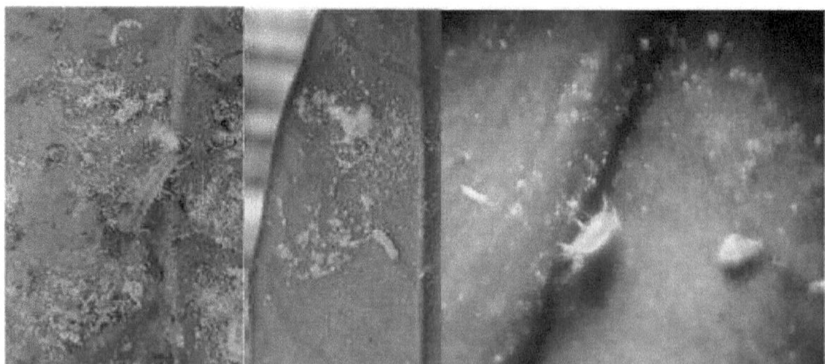

Illustration 44: (Right) Crhysoperla carnea (Crhysoperla carnea) near nests.
(Centre) Larva of Clistostethus arcuatus.
(Left) Mites (Euseius spp.)

These species did not control the population of *Paraleyrodes minei* but did help to reduce it. There were cases of eggs and nymphs in untidy colonies without honeydew. They did not stain the fruit and leaves with honeydew and whitish filaments.

We confirm the results observed by other authors previously that parasitoids are not present in *Paraleyrodes minei* populations (Laccarino, 1989; García, et al., 1992; Argov, 1994; García-Marí, 2018).

Illustration 45: (Right) Pupae of Conwentziapsociformis. (Left) Adult of Conwentziapsociformis.

In contrast, parasitism by *Cales noacki alone* was able to keep the cottony fly *(Aleurothrixus floccosus)* below the damage threshold. This was evident in the lack of progression of generations, little honeydew secretion and numerous exoskeletons with a kite characteristic of *Cales noacki.*

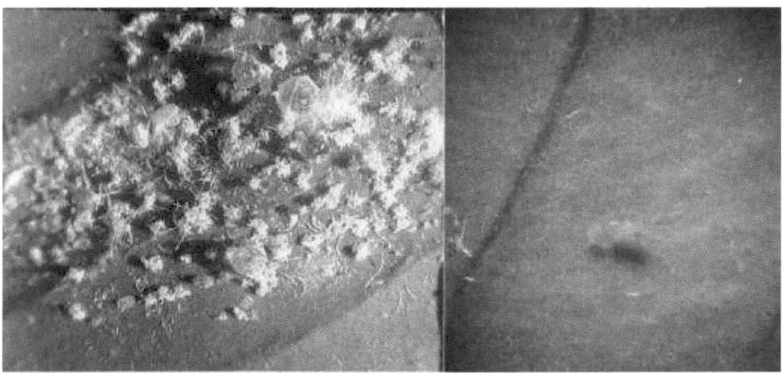

Figure 46: Adults of Cales noacki on colonies of Aleurothrixus floccosus.

Once the phytosanitary treatment was applied, yellow chromatic plates were placed to monitor the auxiliary fauna and how the treatment affected them. A total of 4 traps were taken over a period of 10 weeks with a frequency of two weeks.

A total of 94 natural enemies were counted: 53 parasitoids of *Cales noacki* and 41 predators, which are divided into 25 *Chalcidoidea*, 1 *Ichneumonoidea*, 85 *Hymenoptera*, 1 *Coleoptera* and 8 *Neuroptera* superfamilies.

Figure 11 shows the ratio of adults of the superfamily *Chalcidoidea* expressed as catches per plate, where the catches of *Cales noacki* stand out:

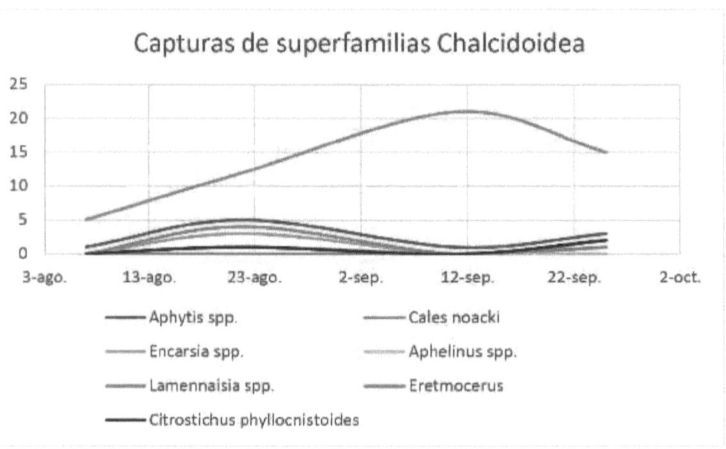

Figure 10: Catches of superfamily Chalcidoidea.

Catch numbers were low and were reflected in the August colour trap assessment.
From 7 August, the first identification and counting of the first chromatic trap was done. The treatment was effective as this was when the smallest values of captures were produced, as well as the whitefly and cottony fly whose severity was at a minimum.

The proportion of the remaining *Chalcidoidea* was much lower than that of *Cales noacki*. In this case, more adult *Cales noacki* were captured with 10 parasitoids per trap per week, while only 6 total *Chalcidoidea* predators were captured.

Chrysopidae carnea, Braconidae and *Clitostethus arcuatus* were found on only one occasion on the plates, but were observed repeatedly in the field, this species considered to be a fly predator, but the fact that only one individual was found could not be considered as such. The flight of these super families was difficult to estimate due to the few captures, but it was thought to start from August affected by the treatment.

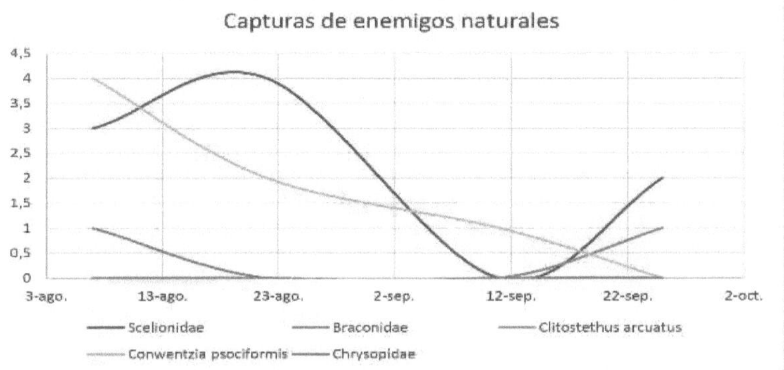

Graph 11: Catch of natural enemies.

Conwentzia psociformis shows greater activity throughout the year due to our sunshine between April and October. The very structure of citrus trees provides greater stability of foliage and results in greater abundance.

The flight of *Scelionidae* started quickly after the treatment, but gradually decreased and ended in September. On the other hand, phytoseiid mites such as *Euseius spp. began* to appear very frequently in the field in September, and with the rains.

In September, natural enemy populations begin to recover, with an increase in their presence on the plates. Finally, the sampling concludes with the recovery of the populations after several adversities that controlled the fly pests.

4. 3. Evaluation of the phytosanitary treatment

The efficacy of the treatment depended on the penetration of the active substance into leaf tissues and translocation within the plant. Penetration and systemic action, and consequently efficacy, could be reduced if applications are made during periods of physiological stress in the crop.

Washing by phytosanitary treatment eliminated adults, newly hatched nymphs, honeydew and part of the nymphs' wool, facilitating the action of natural enemies and consequently increasing biological control.

Paraffin oil is a non-systemic emulsifiable concentrate with contact insecticidal activity for foliar application. The oil cleaned the leaf surface and was effective against whitefly eggs and nymphs immediately. It had a fast action reducing the severity present on the leaves immediately, although it was quickly degraded by the high temperatures in August.

Movento Gold had a double systemic action, i.e. the ability to act upwards and downwards through the sap and therefore be more effective, especially for insects that ingest it, and with a notable effect on the reproductive potential of adults. The persistence of the product against the pest is high, despite having a moderate speed of action because it does not have a shock effect on the pests.

Only a single treatment was used and after its application a decrease in whitefly and cotton fly populations was observed. However, further testing would be needed to determine whether it really has any effect on them and the auxiliary fauna.

Spirotetramat in citrus fruits has a safety period of 21 days from its application. Since Navelinas oranges are an early variety whose cutting takes place from mid-October to the end of January, the phytosanitary action has minimal traces and meets the purchase requirements of supermarkets.

5. Conclusion

The following conclusions can be drawn from this study on the population evolution and control of *Paraleyrodes minei* and *Aleurothixus floccosus* on the farm located in Albatera from June to September 2019:

- Knowledge of the population dynamics of *P. minei and A. floccosus* in the summer season has been broadened to determine the optimal time for phytosanitary application, as well as the proper application of biological control and cultural measures.
- Treatments with Spirotetramat 10% and Paraffin Oil on the farm have reduced populations and stopped the overlapping of generations of both phytophagous pests. However, it also reduced the populations of natural enemies.
- Natural enemies allow control of A. floccosus but not of P. minei. In August, the auxiliary fauna population is low but there is no damaging generation. Subsequently, it recovers and starts to control the pests again.
- *Paraleyrodes minei and Aleurothrixus floccosus* are considered to be controlled by integrated pest management as they do not cause damage to the trees and do not depreciate the fruit.

6. Bibliography

Abad V., 1984. History of the orange. Comité Gestión Gestión Exportación Frutos Cítricos, Valencia, Spain.

Albuquerque Maranháo, E., Albuquerque Maranháo, E., 2009. Entomopathogenic fungi: important tool for the control of whiteflies (homoptera: aleyrodidae), Recife: Anais da Academia Pernambucana de Ciencia Agronómica.

Agustí M., 2003. Citricultura. Ediciones Mundi-Prensa. Madrid, Spain.

Agustí M., Martínez-Fuentes A., Mesejo C., Juan M., Almela V., 2010. Citrus fruit set and development, Mediterranean Agroforestry Institute, Polytechnic University of Valencia, Department of Agriculture, Fisheries and Food.

Argov, 1994. The woolly whitefly, a new pest in Israel. Alon-Hanoteoa, Issue 48, pp. 290-292.

Aznar, J.A., Pérez, J.C., Galdeano, E. 2015. Analysis of the Spanish citrus sector. Cajamar Chair of Economics and Agri-Food of the University of Almeria. http://www.besana.es/sites/default/files/analisis-del-sector-citricola-espanol-2.pdf

Bathon, H. y J. Pietrzik. 1986. On the feeding habits of the archmoth, Clitosthetus arcuatus (Rossi) (Col.,Coccinellidae), an exterminator of the cabbage moth aphid, Aleurodes proletella Linné (Hom., Aleurodidae). J. Appl. Ent. 102, 321-326.

Bellows, Meisenbacher & Headrick, 1998. Field Biology of Paraleyrodes minei (Homoptera: Aleyrodidae) in Southern California. Entomological Society of America, pp. 277-281.

Bravo, J. 2014. Oranges: An export alternative. Ministry of Agriculture Chile. http://www.odepa.cl/wp-content/files mf/1419372212Naranjas2014.pdf

Byrne & Bellows, 1991. Whitefly biology. Ann. Rev. Entomol, Issue 36, pp. 431-457.

Byrne & Bellows, Gill, 2016. Spanish Society of Applied Entomology. [Online] Available at: http://seea.es/index.php/divulgacion/moscas-blancas-de-los-citricos

deBACH, P., 1975. Biological control of insect pests and weeds. Editorial Continental, S.A., pp. 33-34.

Carrero, J. M. 1979a. Contribution to the study of the biology of the white citrus fruit fly, Aleurothrixus floccosus Mask. in the Valencian region. Biological data (Corbera and Gandía), 1975. An. INIA. Ser. Prot. Veg., 9: 107-113.

Carrero, J. M. 1979b. Contribution to the study of the biology of the white citrus fruit fly, Aleurothrixus floccosus Mask. in the Valencian region. III.2. Field biology, Manises, 1975. An. INIA. Ser. Prot. Veg., 9: 115-132.

Carrero, J. M. 1979c. Contribution to the study of the biology of the white citrus fruit fly, Aleurothrixus floccosus Mask. in the Valencian region. V. Studies prior to the establishment of population dynamics. 1. Sampling. An. INIA. Ser. Prot. Veg., 9: 163-175.

Cano, A., Arnao, M.B. 2004. Hydrophilic and lipophilic antioxidant activity and vitamin c content of commercial orange juices: relationship with their organoleptic characteristics. Food Science and Technology 4 (3): 185-189.

Dittrich, V. y G. H. Ernst. 1990. Chemical control and insecticide resistance of whiteflies, pp. 263-286. En D. Gerling (ed.), Whiteflies: their Bionomics, Pest Status and Management. Intercept Ltd, Wimborne, UK.

Dietrich, 2005. Keys to the families of Cicadomorpha and subfamilies and tribes of Cicadellidae (Hemiptera: Auchenorrhyncha). Florida Entomologist, 88(4): 502-517. [En línea] Available at:
http://journals.fcla.edu/flaent/article/view/75474/73132

FAO, 2012. Fresh and processed citrus. Annual statistics 2012. [Online] Available at: http://www.fao.org/fileadmin/templates/est/COMM MARKETS MONITORING/Citrus/Docum ents/CITRUSBULLETIN 2012.pdf

FAO. 2018. FAO Statistics (FAOSTAT). [Online] Available at:
http://www.fao.org/faostat/es/#data/QC

FAO, 2020. Citrus fruits. [Online] Available at:
http://www.fao.org/3/y5143s/y5143s0z.htm#TopOfPage

García, G., Alba, G. & Segura, G., 1992. Presence of Paraleyrodes sp. Pr. Citri (Bondar,1931) (Insecta: Homoptera: Aleyrodidae) in citrus crops in the province of Malaga (southern Spain): Biological and ecological aspects of the pest. Bol. San.Veg. Pests, pp. 3-9.

García Marí F., Costa Comelles J., Ferragut Pérez F., Laborda Cenjor R., 1989. Agricultural Pests I. Mites and exopterygote insects. 1989. Ed. Servicio de publicaciones Universidad Politécnica de Valencia. 269 pages.

García García, J. 2014. Analysis of the lemon tree sector and economic evaluation of its cultivation. IMIDA. Murcian Institute for Agricultural and Food Research and Development. Regional Ministry of Agriculture and Water of the Region of Murcia. Murcia. 148 pp.

García Marí, F. 2009. Field guide. Citrus pests and their natural enemies. Phytoma España. Polytechnic University of Valencia. 176pp.

García Marí, F. 2012. Citrus pests: Integrated Management in Mediterranean climate countries. Polytechnic University of Valencia. 556pp.

García-Marí, 2018. Geographical distribution and seasonal and interannual evolution of the whitefly Paraleyrodes minei (Hemiptera: Aleyrodidae) in citrus crops in the eastern Iberian Peninsula. Levante Agírcola, p. 41.

Garcia Marí, F., 2018. Paraleyrodes dynamics. Levante Agrícola, pp. 38-39.

García-Marí, 2019. Course on Sampling, Identification and Management of Pests and their Natural Enemies in Citrus and Persimmon. Phytoma Spain. Polytechnic University of Valencia.

Garrido, Tarancón, Busto, d. & Martínez, 1977. Expansion of Cales noacki How. from a point release and larval stages of Aleurothrixus Floccosus Mask preferred by the parasite. Ser. Prot. Veg, Volume 7, pp. 145-175.

Garrido, 1983. Citrus whiteflies in Spain. Levante Agrícola, pp. 27-34.

Garrido, 1989. Citrus whitefly (Aleurothrixus floccosus Mask). El campo. Boletín de información agraria, Issue 113, pp. 42-46.

Garrido, 1991. Aleuródidos de los cítricos españoles. Levante Agrícola, Volume 1st quarter, pp. 4453.

Garrido, 1992a. Considerations and problems of aleurodids in citrus. Phytoma España, Issue 40, pp. 129-137.

Garrido, 1992b. Current status of whitefly in Spanish citrus and guidelines for its control. Levante Agrícola, Issue 9, pp. 157-167.

Garrido, 1994b. Current problems of whiteflies in citrus cultivation (II). Phytoma, Issue 59, pp. 12-24.

Garrido, 1994c. Biological control of whitefly. In: I. congreso de citricultura de la plana. s.l.:s.n., pp. 243-267.

Gerling, 1990. Natural enemies of whiteflies: predators and parasitoids. Intercept Ltd. Wimborne, pp. 147-186.

González-Sicilia E., 1968. El cultivo de los agrios, Ed. Bello, Valencia, Spain.

Good Agricultural Practices (GLOBALG.A.P.), 2015. GLOBALG.A.P. Certification. Requirements for producers. [Online] Available at: http://www.globalgap.org/es/what-we-do/globalg.a.p.- certification/

Gold, C. y Altieri, M. 1989. The effects of intercropping and mixed varieties of predators and parasitoids of cassava whiteflies (Hemiptera: Aleyrodidae) in Colombia. Bull. Ent. Res., 79: 115121.

Halsey & Mount, 1978. Whitefly of the world. A systematic catalogue of the aleyrodidae (Homoptera) with host plant and natural enemy data. *British Museum (Natural history) y John Wiley and Sons, Great Britain.*

Heinz, K. M., J. R. Brazzle, C. H. Picket, E. T. Natwick, J. M. Nelson y M. P. Parrella. 1994. Predatory beetle may suppress silverleaf whitefly. California Agriculture, 48: 35-40.

Hoelmer, Kim & Osborne, Lance & Bennett, Fred & Yokomi, Ray. (1994). Biological control of sweetpotato whitefly in Florida.

Kimball, D.A., 2002. Citrus processing. Ed. Acribia, Spain.

Laccarino, 1989. Descrizione di Paraleyrodes minei n. sp. (homoptera:

Aleyrodidae), nuevo aleirodidae degli agrumi, in Siria, Napoli: s.n.

Leon M. G and Kondo T. 2017. Citrus insects and mites: Illustrated compendium of harmful and beneficial species, with techniques for integrated pest management. New agricultural knowledge collection. Corpoica Editorial. 1-183 pp.

Llorens, 1994. Introduction, biology and control of the citrus whitefly Dialeurodes citri (homoptera, Aleyrodidae) in the province of Alicante. Polytechnic University of Valencia.

Loi, G. 1978. Eco-ethological observations on the coccinellidae scimnino beetle Clitostethus arcuatus (Rossi) predador of Dialeurodes citri (Ashm.) in Tuscany. Frustula Entomologica, 1: 123-145

Lucas Espadas, A. 2009. Citrus pests and diseases in the Region of Murcia. Department of Agriculture and Water. Murcia. 64 pp.

Luppichini, Paola & Ripa, Renato & Pilar, S & Fernando, Larral & Inia, Rodríguez & Cruz, La & Larral, Pilar. (2007). Integrated management of cotton whitefly in citrus.

MAGRAMA, 2015. Analysis of the citrus fruit sector during the 2014-2015 season. Ministry of Agriculture, Food and Environment. [Online] Available at: http://www.magrama.gob.es/es/prensa/noticias/el-ministerio-de-agricultura-alimentaci%C3%B3n-y-medio-ambiente-analiza-analiza-con-el-sector-de-los-c%C3%ADtricos-el-curso-de-la-campa%C3%B1a-20142015/tcm7-363237-16

Malausa, J. C., E. Franco y P. Brun. 1988. Aclimatization on the French Riviera and in Corsica of Serangium parcesetosum (Col.:Coccinellidae) predator of the citrus whitefly, Dialeurodes citri (Hom.:Aleyrodidae). Entomophaga, 33: 517-519.

Mansilla Vázquez P., Salinero Corral C., Pérez Otero R., Iglesias Vázquez C., 2003. Environmental techniques of rationalisation in the use of chemical products, symptoms, monitoring and control of phytopathogens of the most frequent crops in Galicia. Xunta de Galicia. Conselleria de Agricultura, Ganadería y Política Agroalimentaria. [Online] Available at: http://mediorural.xunta.gal/fileadmin/arquivos/publicacions/Agricultura/medidasagroambie ntales.pdf

MAP. 2018. Agrarian Statistics Yearbook. Ministry of Agriculture, Food and Environment. Accessed June 2020 [Online] Available at:

https://www.mapa.gob.es/es/estadistica/temas/publicaciones/anuario-de-statistics/2018/default.aspx?part=3&chapter=07&group=8

MAPA 2019. Agricultural Statistics Yearbook. Ministry of Agriculture, Food and Environment. Accessed June 2020. [Online] Available at: https://www.mapa.gob.es/es/estadistica/temas/publicaciones/anuario-de-estadistica/2019/default.aspx?parte=3&capitulo=07&grupo=8

MAPA, 2020. Register of phytosanitary products. Ministry of Agriculture, Food and Environment. [Online] Available at: http://www.mapama.gob.es/es/agricultura/temas/sanidad-vegetal/productos-fitosanitarios/registro/menu.asp

MAPAMA. 2014. Guía de Gestión Integrada de Plagas-Cítricos. Ministry of Agriculture, Fisheries and Food. [Online] Available at: http://www.mapama.gob.es/es/agricultura/temas/sanidad-vegetal/GUIACITRICOStcm30- 57942.pdf

MAPAMA. 2014. Guía de Gestión Integrada De Plagas-Cítricos. Ministry of Agriculture, Fisheries and Food. [Online] Available at: http://www.mapama.gob.es/es/agricultura/temas/sanidad-vegetal/GUIACITRICOStcm30- 57942.pdf

Maroto J.V, 1998. Historia de la agronomía Ed. Mundi-Prensa, Madrid, Spain.

Ministry of Agriculture, Fisheries and Food. Tuesday, 19 November 2019. Retrieved 12/05/2020: https://www.lamoncloa.gob.es/espana/eh18-19/agricultura/Paginas/agriculturayganaderia.aspx

Mound, 1973. Spanish Society of Applied Entomology. [Online] Available at: http://seea.es/index.php/divulgacion/moscas-blancas-de-los-citricos

Mount & Halsey, 1978. Whitefly of the world. A systematic catalogue of the aleyrodidae (Homoptera) with host plant and natural enemy data. British Museum (Natural history) y John Wiley and Sons, Great Britain.

Monera Olmos R.V., 1999. Valencian citrus fruits: pure vitamin, source of health, rural life. Ministry of Agriculture, Food and Environment. P.: 26-29. [Online] Available at: http://www.magrama.gob.es/ministerio/pags/biblioteca/revistas/pdf vrural/Vrural 1999 85 26 29.pdf

Onillon, J. C. 1969. A propos de la presence en France d'une nouvelle espèce d'aleurode nuisible aux citrus, Aleurothrixus floccosus Maskell (Homopt. Aleurodidae). C. R. Acad. Agri. France, 55: 937-941.

Onillon, J. C. 1973. Possibilities of regulating populations of Aleurothrixus floccosus Mask. (Homopt.,Aleurodidae) on citrus by Cales noacki How. (Hymenopt., Aphelinidae). EPPO Bulletin, 3: 17-23.

Onillon, 1977. Aspects of the ecology of some aleurodids. Bol. Pest Serv., Issue 3, pp. 175- 198.

Onillon, J. C. and J. Onillon. 1972. Contribution a l'étude de la dynamique des populations d'homoptères inféodés aux agrumes. III. Introduction, in the Alpes-Maritimes of Cales noacki How (Hymenopt., Aphelinidae), parasite of Aleurothrixus floccosus Mask. (Homopt., Aleurodidae). French Academy of Agriculture. Extract from the minutes of the Session of 15 March: 365-370.

Osman, I. G. 1971. The whitefly Eggs and First Larval Stages as Prey for certain Phytoseiid Mites. Rev. Zool. Bot. Afr., 84: 79-82.

Santaballa, E., C. Borras and P. Colomer. 1980. Control of the citrus whitefly Aleurothrixus floccosus Mask. Bol. Pest Serv. 6:109-118.

Soto, A., F. Ohlenschlager and F. Garcia-Marí. 1999. Status of biological control of citrus whiteflies Aleurothrixus floccosus, Parabemisia myricae and Dialeurodes citri in Valencia. Levante Agrícola, 4th quarter: 475-484.

Soto, A., F. Ohlenschlager and F. Garcia-Marí. 2001. Population dynamics and biological control of whiteflies Aleurothrixus floccosus, Dialeurodes citri and Parabemisia myricae (Homoptera: Aleyrodidae) in Valencian citrus. Bol.San. Veg. Pests, 27: 3-20.

Soto, García-Marí & Ohlenschláger, 2001. Population dynamics and biological control of the whiteflies Aleurothrixus floccosus, Dialeurodes citri and Parabemisia myricae (Homoptera: Aleyrodidae) in Valencian citrus. Boletin de sanidad vegetal y plagas, Issue 3, pp. 3-20.

Soler, García-Marí & Alonso, 2002. Seasonal evolution of auxiliary entomofauna. Boletín de Sanidad Vegetal de Plagas, pp. 133-149.

Soto, García-Marí, 2016. Spanish Society of Applied Entomology. Available at: http://seea.es/index.php/divulgacion/moscas-blancas-de-los-citricos

Soler, J. 2006 Citrus. Varieties and cultivation techniques. Ediciones Mundi-Prensa.

Soto, 1999. Population dynamics and biological control of the citrus whiteflies Parabemisi myricae (Kuwana), Aleurothrixus floccosus (Maskell) and Dialeurodes citri (Ashmead) (Homoptera: Aleyrodidae). Universidad Politécnica de Valencia Escuela Técnica Superior de Ingenieros Agrónomos: s.n.

Soto Sánchez, A. & Langa Llueca, E., 2018. Biology and seasonal dynamics of Paraleyrodes minei. Universitat Politècnica de Valéncia.

Ragusa Ch., S., H. Tsolakis y A. Sinacori. 1991. Observations on the postembryonic development and oviposition of three species of phytoseeds in the presence of Parabemisia myricae (Kuwana). Convegno su Attività del Gruppo di Ricerca "Lotta biologica", Protureetto finalizzato M.A.F. "Lotta biologica e integrata per la Difesa delle Colture agrarie e delle Piante forestali", Acireale Noviembre, 151-155.

Teich, Y. 1966. Mites of the family Phitoseiidae as predators of the tobacco whitefly, Bemisia tabaci Gennadius. Israel J. Agric. Res., 16: 141-142.

Thomas H., 2001. World Citrus Production and Consumption Projections for 2010 from the Food and Agriculture Organization of the United Nations (FAO), United States. [Online] Available at: http://www.fao.org/3/a-x6732s/x6732s03.pdf

IVIA, 2015. Integrated management of pests and diseases in citrus. Valencian Institute of Agricultural Research. [Online] Available at: http://gipcitricos.ivia.es/citricultura- valenciana

IVIA, 2019. Mediterranean Horticulture. Instituto Valenciano de Investigaciones Agrarias. [Online] Available at: http://www.horticulturablog.com/2013/04/la-industria-de-los-citricos- espanoles.html

IVIA, 2020. Integrated Pest and Disease Management in Citrus. Pests and diseases. Valencian Institute of Agricultural Research. [Online] Available at: http://gipcitricos.ivia.es/area/plagas-principales/moscas-blancas

SIAR Network, 2020. Agrometeorological Report of the Orihuela-La Murada station. Accessed June 2020. [Online] Available at: http://riegos.ivia.es/datos-meteorologicos

Zaragoza S., Pina Lorca J.A., Forner M.A., Navarro L., Medina A., Soler G., Chomé Fuster P.M., 2011. Citrus varieties. Ministerio de Medio Ambiente y Medio Rural y Marino, Spain. [Online] Available at: http://www.magrama.gob.es/es/ministerio/servicios/publicaciones/Variedad es de Citricosp rimeras p%C3%A1ginastcm7-212147.pdf. Citrus varieties.

Zaragoza, 1993. Pasado y presente de la citricultura española. Generalitat Valenciana, Serie Div. Téc. no 8, Valencia, Spain.

Wysoki, M. y M. Cohen. 1983. Mites of the family Phytoseiidae (Acarina, Mesostigmata) as predators of the Japanese bayberry whitefly, Parabemisia myricae Kuwana (Hom., Aleyrodidae. Agronomie, 3: 823-825

Ferra García-Marí (2009). Geographical distribution and seasonal and interannual evolution of the whitefly Paraleyrodes minei (Hemiptera: Aleyrodidae) in citrus crops in the east of the Iberian Peninsula. International Citrus Journal Levante agrícola. 2nd Quarter 2018. No: 441.

Langa Llueca, Eduardo (2018). Biology and seasonal dynamics of Paraleyrodes minei from Universidad Politécnica de Valencia.

7. Annexes

7. 1. Behaviour of stocks in relation to climatic factors

In this annex we have proceeded to analyse what repercussion this factor may have on the populations of the whitefly *(Paraleyrodes mieni)* and the cotton fly *(Aleurothrixus floccosus).* Looking at the dynamics of both types of flies, we can see that graphs 14 and 15 show different behaviours throughout the sampling period.

According to the SIAR of the Albatera area, the climate is typical of a semi-arid Mediterranean climate. The climate is temperate, during the course of the year, the temperature generally varies from 5°C to 32°C and rarely drops below 1°C or rises above 35°C.

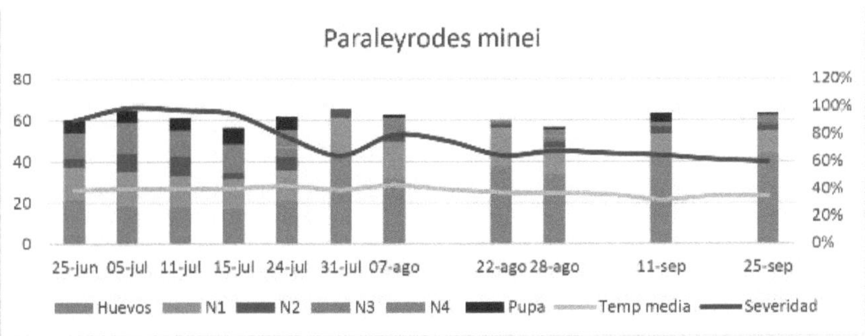

Figure 12: Mean temperature, population distribution and severity of Paraleyrodes minei.

As mentioned above, the average threshold temperature is 7.5oC, which means that throughout the period the pest evolves rapidly due to temperatures above 30°C in the months of June and July. As shown in Graph 14, the period of maximum temperature coincides with the period of maximum egg density and N1 with a severity of more than 60%. There are overlapping generations of *Paraleyrodes minei* whose proportions are distributed proportionally to their stage, but with a high severity.

After treatment, the natural enemies are affected and, together with high temperatures, the adults begin to lay their eggs exponentially, peaking on 7 August. After the temperature peak, the temperature gradually decreases, as does the severity.

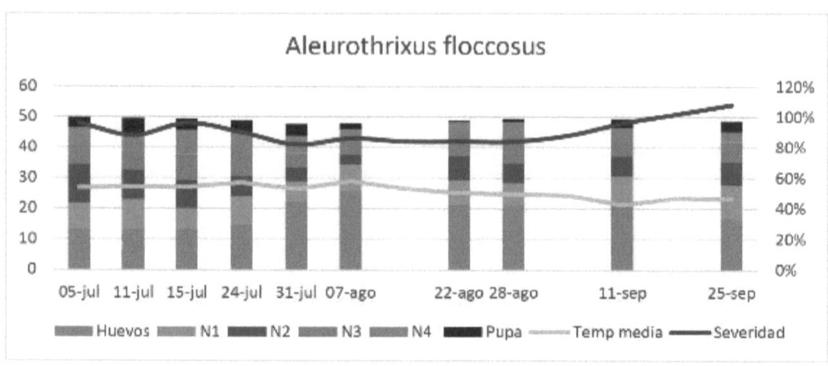

Graph 13: Mean temperature, population distribution and severity of Aleurothrixus floccosus.

Comparing both figures we can see that in the cottony fly there is a greater severity with the presence of nymphs and eggs in spite of the lower temperatures at the end of the sampling. This is possibly due to the treatment and the scarce auxiliary fauna allowing more eggs to be laid on the leaves, which has also led to a higher number of nymphs. Apart from this detail, it is logical to find no differences in the phenology of *Paraleyrodes minei and Aleurothrixus floccosus* due to the different meteorological factors.

The weather conditions in September 2019 were favourable for the auxiliary fauna, as numerous disorderly colonies began to appear and a greater number and diversity of catches, once again generating biological control.

Continuing with the climatic factor of rainfall, Graph 16 shows that, in the year 2019, the maximum volume of water recorded was on 9 September with a rainfall of 291 mm. The average annual rainfall is 13 mm/day, with September and October being the wettest months, and July and August the driest.

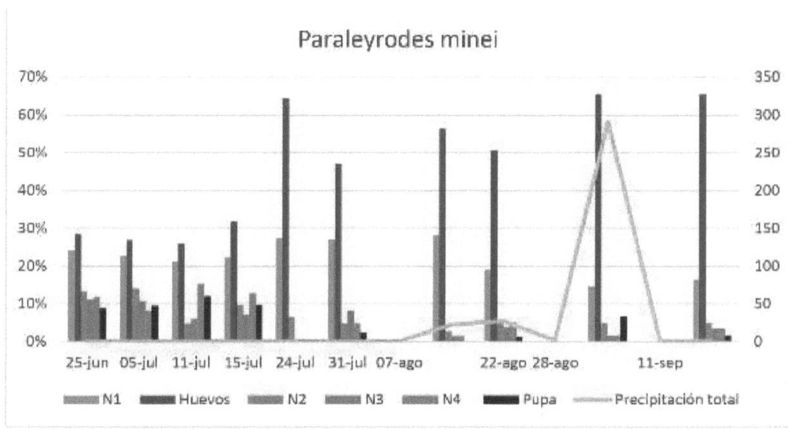

Graph 14: Daily precipitation, in mm, during the period of the work.

In the months of August and September, the density of eggs increases with respect to the other stages, expecting a new overlapping after the treatment. According to the results obtained, the rains in August did not have a knock-on effect on the populations, affecting only the colonies located at the edges of the leaves. However, the torrential rain episode (DANA) reduced the population density to the control point of the study plots.

The heavy rains washed the trees like the treatment, removing the colonies from the leaves, leaving few N1 nymphs and producing a new clutch of eggs, which increases the proportionality of this to the other stages.

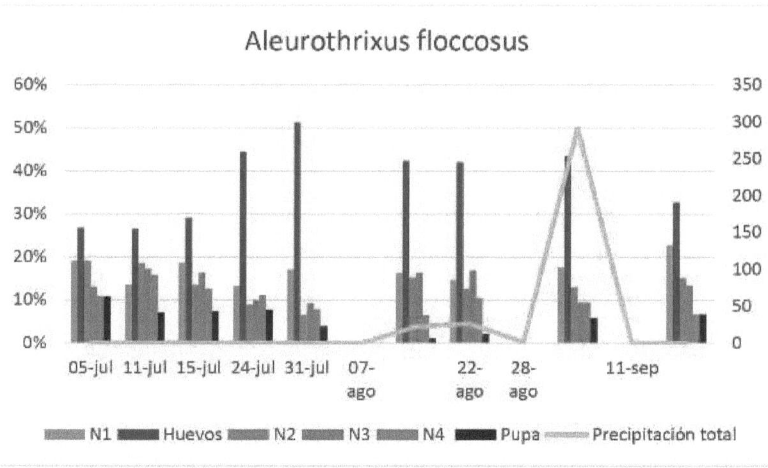

The observed population is quite low and this is possibly the main reason why these changes are not detected. Therefore, in order to confirm the effect of rainfall on *P. minei and A. floccosus* populations, *it* would be necessary in the future to extend the studies with more abundant populations in which this factor can be observed more significantly.

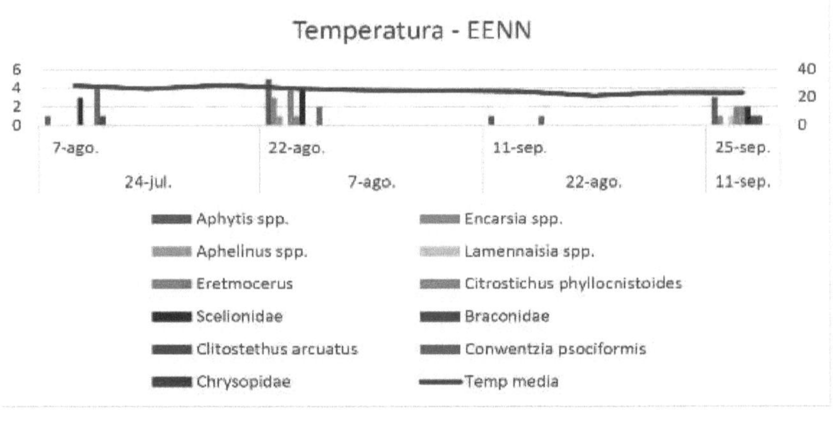

Despite the high temperatures in July, the auxiliary fauna is at levels capable of preying on and parasitising the flies. In August and early September, on the other hand, populations are at a minimum.

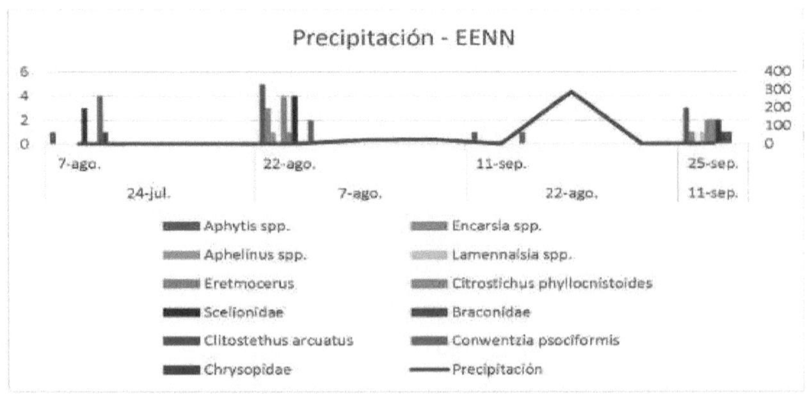

With the progressive drop in temperature and the rains, they favour the autochthonous appearance of auxiliary fauna by observing them in the field and the capture plate. Climatic conditions favour natural enemies such as *Aphytis spp., Eretmocerus and Scelionidae.*

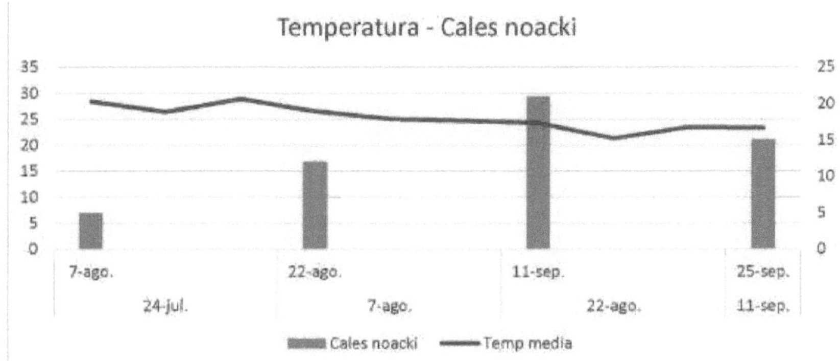

In addition, the weather conditions in July, with a slight drop in temperatures, allow a rebound in the Cales noacki population at a time when the pest tends to grow exponentially and generate an efficient control with the return of disordered colonies without honeydew.

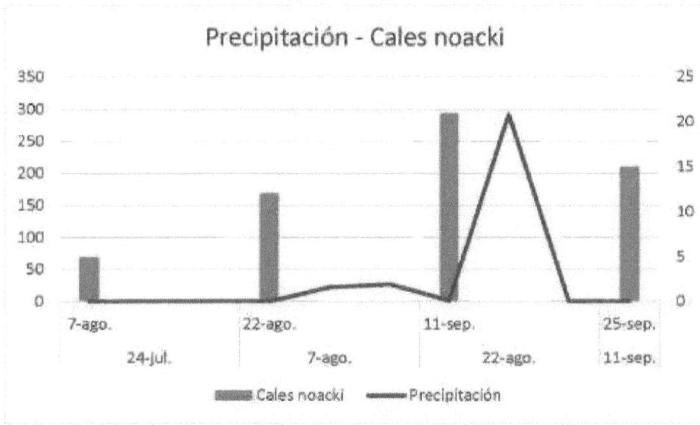

From the first sampling of natural enemies it is increasing until the episode of torrential rain in September 2019 (DANA) where it decreases or they are not captured in that period except for *Cales noacki*. In the last plate, captures of *Cales noacki* decrease because the presence of *Aleuthrixus floccosus* decreases with the rain wash in mid-September and they do not have as many nymphs to parasitise.

Although a relationship between population dynamics and temperature can be observed, no conclusions can be drawn from the results obtained and the observation period used to carry out the work.

7. 2. Table of records of the superfamily Chalcidoidea
The following tables show the total of each superfamily and species, as well as the evolution of adults captured on the yellow chromatic plates in Navelina Orange. The Hymenoptera families and Chalcidoidea superfamilies of captured (table 9):

Sample	Placement date	Date of collection	number of days	Trap (1 or 2)	Aphytis spp.	Cales noacki	Encarsia spp.	Aphelinus spp.	Lamennaisia spp.	Citrostichus phyllocnistoides
1	24-Jul.	7-Aug.		1	1	5	0	0	0	0
	7-Aug.	22-Aug.		1	5			1	0	1
	22-Aug.	11-Sep.		1	1	21	0	0	0	0
	11-Sep.	25-Sep.		1			1		1	
				Total				1	1	

Table 9: Abundance of the superfamily Chalcidoidea and species caught in the traps

7. 3. Predator register table

Superfamily:				Platygastroidea	Ichneumonoidea	Coccinellidae	Coniopterygidae	Chrysopidae
Placement date	Date of collection	number of days	Tra mpa	Scelionidee	Braconidae	Clitostethus arcuatus	Conwentzia psociformis	Chrysopidae
24-Jul.	7-Aug.		1		0	0		1
7-Aug.	22-Aug.				0	0		0
22-Aug.	11-Sep.			0	0	0	1	0
11-Sep.	25-Sep.				1	1	0	0
Total					1	1		1

7.4. Template identifies and accounts for whiteflies

Simplified Whitefly Sampling										
Whitefly species:										
Finca:										
Variety										
Sample size: 40 sheets. 1 sheet /ar			bowl							
Date										
	Eggs	N1	N2	N3	N4	Pupa	Adults	Severity		
1										
5										

8

77

21

25

26

29

30

31

35											
Conta r	0	0	0	0	0	0	0				
TOTAL	#DIV/0!	#DIV/0!	#DIV/0!	#DIV/0!	#DIV/0!	#DIV/0!	#DIV/0!	0			
								#DIV/0!				
Sumatorio H+N1	#DIV/0!			Mean value severity scale:				0			

Printed by Books on Demand GmbH, Norderstedt / Germany